Calculus
Essentials
FOR
DUMMIES®

by Mark Ryan

WILEY

Wiley Publishing, Inc.

Calculus Essentials For Dummies®

Published by
Wiley Publishing, Inc.
111 River St.
Hoboken, NJ 07030-5774
www.wiley.com

Copyright © 2010 by Wiley Publishing, Inc., Indianapolis, Indiana

Published simultaneously in Canada

For general information on our other products and services, please contact our Customer Care Department within the U.S. at 877-762-2974, outside the U.S. at 317-572-3993, or fax 317-572-4002.

For technical support, please visit www.wiley.com/techsupport.

Wiley also publishes its books in a variety of electronic formats. Some content that appears in print may not be available in electronic books.

Library of Congress Control Number: 2010924588

ISBN: 978-0-470-61835-6

Manufactured in the United States of America

10 9 8 7 6 5 4

WILEY

About the Author

Mark Ryan, a graduate of Brown University and the University of Wisconsin Law School, has been teaching math since 1989. He runs the Math Center in Winnetka, Illinois (www.themath center.com), where he teaches high school math courses including an introduction to calculus and a workshop for parents based on a program he developed, *The 10 Habits of Highly Successful Math Students*. He also does extensive one-to-one tutoring for all levels of mathematics and for standardized test preparation. In high school he twice scored a perfect 800 on the math portion of the SAT, and he not only knows mathematics, he has a gift for explaining it in plain English. He practiced law for four years before deciding he should do something he enjoys and use his natural talent for mathematics. Ryan is a member of the Authors Guild and the National Council of Teachers of Mathematics.

Calculus Essentials For Dummies is Ryan's sixth book. *Everyday Math for Everyday Life* was published in 2002, *Calculus For Dummies* (Wiley) in 2003, *Calculus Workbook For Dummies* (Wiley) in 2005, *Geometry Workbook For Dummies* (Wiley) in 2007, and *Geometry For Dummies,* 2nd Ed. (Wiley) in 2008. His math books have sold over a quarter of a million copies.

Publisher's Acknowledgments

We're proud of this book; please send us your comments at http://dummies.custhelp.com. For other comments, please contact our Customer Care Department within the U.S. at 877-762-2974, outside the U.S. at 317-572-3993, or fax 317-572-4002.

Some of the people who helped bring this book to market include the following:

Acquisitions, Editorial, and Media Development

Project Editor: Corbin Collins

Senior Acquisitions Editor: Lindsay Sandman Lefevere

Copy Editor: Corbin Collins

Assistant Editor: Erin Calligan Mooney

Editorial Program Coordinator: Joe Niesen

Technical Editors: Eric Boucher, Jon Lark-Kim

Senior Editorial Manager: Jennifer Ehrlich

Editorial Supervisor and Reprint Editor: Carmen Krikorian

Editorial Assistants: Rachelle Amick, Jennette ElNaggar

Senior Editorial Assistant: David Lutton

Cartoon: Rich Tennant (www.the5thwave.com)

Composition Services

Project Coordinator: Sheree Montgomery

Layout and Graphics: Carl Byers, Carrie A. Cesavice, Mark Pinto, Melissa K. Smith

Proofreaders: Melissa Cossell, Henry Lazarek

Indexer: Potomac Indexing, LLC

Publishing and Editorial for Consumer Dummies

Diane Graves Steele, Vice President and Publisher, Consumer Dummies

Kristin Ferguson-Wagstaffe, Product Development Director, Consumer Dummies

Ensley Eikenburg, Associate Publisher, Travel

Kelly Regan, Editorial Director, Travel

Publishing for Technology Dummies

Andy Cummings, Vice President and Publisher, Dummies Technology/General User

Composition Services

Debbie Stailey, Director of Composition Services

Contents at a Glance

Table of Contents

Introduction

The mere thought of having to take a required calculus course is enough to make legions of students break out in a cold sweat. Others who have no intention of ever studying the subject have this notion that calculus is impossibly difficult unless you happen to be a direct descendant of Einstein.

Well, I'm here to tell you that you *can* master calculus. It's not nearly as tough as its mystique would lead you to think. Much of calculus is really just very advanced algebra, geometry, and trig. It builds upon and is a logical extension of those subjects. If you can do algebra, geometry, and trig, you can do calculus. Read this jargon-free book, get a handle on calculus, and join the happy few who can proudly say, "Calculus? Oh, sure, I know calculus. It's no big deal."

About This Book

Calculus Essentials For Dummies is intended for three groups of readers: students taking their first calculus course, students who need to brush up on their calculus to prepare for other studies, and adults of all ages who'd like a good introduction to the subject. For those who'd like a fuller treatment of the subject, check out *Calculus For Dummies*.

If you're enrolled in a calculus course and you find your textbook less than crystal clear, *Calculus Essentials For Dummies* is the book for you. It covers the two most important topics in the first year of calculus: differentiation and integration.

If you've had elementary calculus, but it's been a couple of years and you want to review the concepts to prepare for, say, some graduate program, *Calculus Essentials For Dummies* will give you a quick, no-nonsense refresher course.

Nonstudent readers will find the book's exposition clear and accessible. *Calculus Essentials For Dummies* takes calculus out of the ivory tower and brings it down to earth.

This is a user-friendly math book. Whenever possible, I explain the calculus concepts by showing you connections between the calculus ideas and easier ideas from algebra and geometry. I then show you how the calculus concepts work in concrete examples. Only later do I give you the fancy calculus formulas. All explanations are in plain English, not math-speak.

Conventions Used in This Book

The following conventions keep the text consistent and oh-so-easy to follow.

- ✔ Variables are in *italics*.

- ✔ Calculus terms are italicized and defined when they first appear in the text.

- ✔ In the step-by-step problem-solving methods, the general action you need to take is in **bold**, followed by the specifics of the particular problem.

Foolish Assumptions

Call me crazy, but I assume . . .

- ✔ You know at least the basics of algebra, geometry, and trig.

  If you're rusty, you might want to brush up a bit on these pre-calculus topics. Actually, if you're not currently taking a calculus course, and you're reading this book just to satisfy a general curiosity about calculus, you can get a good conceptual picture of the subject without the nitty-gritty details of algebra, geometry, and trig. But you won't, in that case, be able to follow all the problem solutions. In short, without the pre-calculus stuff, you can see the calculus forest, but not the trees. If you are enrolled in a calculus course, you've got no choice — you've got to know the trees.

- ✔ You're willing to do some w_ _ _ .

  No, not the dreaded w-word! Yes, that's w-o-r-k, work. I've tried to make this material as accessible as possible, but it is calculus after all. You can't learn calculus by just listening to a tape in your car or taking a pill — not yet anyway.

Icons Used in This Book

Keep your eyes on the icons:

Next to this icon are the essential calculus rules, definitions, and formulas you should definitely know.

These are things you need to know from algebra, geometry, or trig, or things you should recall from earlier in the book.

The bull's-eye icon appears next to things that will make your life easier. Take note.

This icon highlights common calculus mistakes. Take heed.

In contrast to the Critical Calculus Concepts, you generally don't need to memorize the fancy-pants formulas next to this icon unless your calc teacher insists.

Where to Go from Here

Why, Chapter 1, of course, if you want to start at the beginning. If you already have some background in calculus or just need a refresher course in one area or another, then feel free to skip around. Use the table of contents and index to find what you're looking for. If all goes well, in a half a year or so, you'll be able to check calculus off your bucket list:

___ Run a marathon

___ Go skydiving

___ Write a book

_x_ Learn calculus

___ Swim the English Channel

___ Cure cancer

___ Write a symphony

___ Pull an inverted 360° at the X-Games

For the rest of your list, you're on your own.

The 5th Wave

By Rich Tennant

"What exactly are we saying here?"

Chapter 1

Calculus: No Big Deal

* *

In This Chapter

▶ Calculus — it's just souped-up regular math

▶ Zooming in is the key

▶ Delving into the derivative: It's a rate or a slope

▶ Investigating the integral — addition for experts

* *

*I*n this chapter, I answer the question "What is calculus?" in plain English and give you real-world examples of how it's used. Then I introduce you to the two big ideas in calculus: differentiation and integration. Finally, I explain why calculus works. After reading this chapter, you *will* understand what calculus is all about.

So What Is Calculus Already?

Calculus is basically just very advanced algebra and geometry. In one sense, it's not even a new subject — it takes the ordinary rules of algebra and geometry and tweaks them so that they can be used on more complicated problems. (The rub, of course, is that darn *other* sense in which it *is* a new and more difficult subject.)

Look at Figure 1-1. On the left is a man pushing a crate up a straight incline. On the right, the man is pushing the same crate up a curving incline. The problem, in both cases, is to determine the amount of energy required to push the crate to the top. You can do the problem on the left with regular math. For the one on the right, you need calculus (if you don't know the physics shortcuts).

For the straight incline, the man pushes with an *unchanging* force, and the crate goes up the incline at an *unchanging* speed. With some simple physics formulas and regular math (including algebra and trig), you can compute how many calories of energy are required to push the crate up the incline. Note that the amount of energy expended each second remains the same.

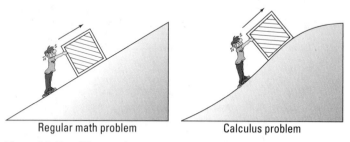

Regular math problem   Calculus problem

Figure 1-1: The difference between regular math and calculus: In a word, it's the curve.

For the curving incline, on the other hand, things are constantly *changing*. The steepness of the incline is *changing* — and it's not like it's one steepness for the first 3 feet and then a different steepness for the next 3 — it's *constantly changing*. And the man pushes with a *constantly changing* force — the steeper the incline, the harder the push. As a result, the amount of energy expended is also *changing*, not just every second or thousandth of a second, but *constantly*, from one moment to the next. That's what makes it a calculus problem. It should come as no surprise to you, then, that calculus is called "the mathematics of change." Calculus takes the regular rules of math and applies them to fluid, evolving problems.

For the curving incline problem, the physics formulas remain the same, and the algebra and trig you use stay the same. The difference is that — in contrast to the straight incline problem, which you can sort of do in a single shot — you've got to break up the curving incline problem into small chunks and do each chunk separately. Figure 1-2 shows a small portion of the curving incline blown up to several times its size.

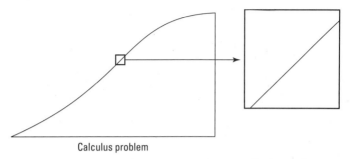

Calculus problem

Figure 1-2: Zooming in on the curve — voilà, it's straight (almost).

When you zoom in far enough, the small length of the curving incline becomes practically straight. Then you can solve that small chunk just like the straight incline problem. Each small chunk can be solved the same way, and then you just add up all the chunks.

That's calculus in a nutshell. It takes a problem that can't be done with regular math because things are constantly changing — the changing quantities show up on a graph as curves — it zooms in on the curve till it becomes straight, and then it finishes off the problem with regular math.

What makes calculus such a fantastic achievement is that it does what seems impossible: it zooms in *infinitely*. As a matter of fact, everything in calculus involves infinity in one way or another, because if something is constantly changing, it's changing infinitely often from each infinitesimal moment to the next.

Real-World Examples of Calculus

So, with regular math you can do the straight incline problem; with calculus you can do the curving incline problem. With regular math you can determine the length of a buried cable that runs diagonally from one corner of a park to the other (remember the Pythagorean Theorem?). With calculus you can determine the length of a cable hung between two towers that has the shape of a *catenary* (which is different, by the way, from a simple circular arc or a parabola). Knowing the exact length is of obvious importance to a power company planning hundreds of miles of new electric cable.

You can calculate the area of the flat roof of a home with regular math. With calculus you can compute the area of a complicated, nonspherical shape like the dome of the Minneapolis Metrodome. Architects need to know the dome's area to determine the cost of materials and to figure the weight of the dome (with and without snow on it). The weight, of course, is needed for planning the strength of the supporting structure.

With regular math and simple physics, you can calculate how much a quarterback must lead a pass receiver if the receiver runs in a *straight* line and at a *constant* speed. But when NASA, in 1975, calculated the necessary "lead" for aiming the Viking I at Mars, it needed calculus because both the Earth and Mars travel on *elliptical* orbits, and the speeds of both are *constantly changing* — not to mention the fact that on its way to Mars, the spacecraft was affected by the different and *constantly changing* gravitational pulls of the Earth, moon, Mars, the sun. See Figure 1-3.

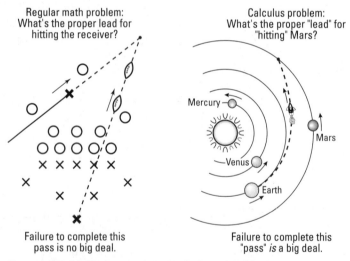

Regular math problem:
What's the proper lead for
hitting the receiver?

Calculus problem:
What's the proper "lead" for
"hitting" Mars?

Mercury

Venus

Earth

Mars

Failure to complete this
pass is no big deal.

Failure to complete this
"pass" *is* a big deal.

Figure 1-3: B.C.E. (Before the Calculus Era) and C.E. (the Calculus Era).

Differentiation

Differentiation is the first big idea in calculus. It's the process of finding a *derivative* of a curve. And a derivative is just the fancy calculus term for a curve's slope or steepness.

In algebra, you learned the slope of a line is equal to the ratio of the *rise* to the *run*. In other words, $Slope = \frac{rise}{run}$. In Figure 1-4, the *rise* is half as long as the *run*, so segment AB has a slope of 1/2. On a curve, the slope is constantly *changing*, so you need calculus to determine its slope.

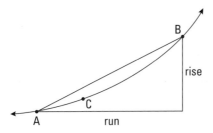

Figure 1-4: Calculating the slope of a curve isn't as simple as *rise* over *run*.

The slope of segment AB is the same at every point from A to B. But the steepness of the curve is changing between A and B. At A, the curve is less steep than the segment, and at B the curve is steeper than the segment. So what do you do if you want the exact slope at, say, point C? You just zoom in. See Figure 1-5.

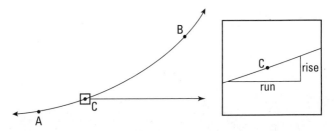

Figure 1-5: Zooming in on the curve.

When you zoom in far enough — actually infinitely far — the little piece of the curve becomes straight, and you can figure the slope the old-fashioned way. That's how differentiation works.

Integration

Integration, the second big idea in calculus, is basically just fancy addition. Integration is the process of cutting up an area into tiny sections, figuring out their areas, and then adding them up to get the whole area. Figure 1-6 shows two area problems — one that you can do with geometry and one where you need calculus.

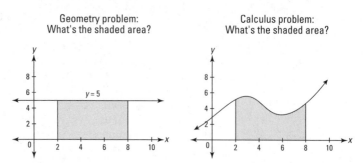

Figure 1-6: If you can't determine the area on the left, hang up your calculator.

The shaded area on the left is a simple rectangle, so its area, of course, equals length times width. But you can't figure the area on the right with regular geometry because there's no area formula for this funny shape. So what do you do? Why, zoom in, of course. Figure 1-7 shows the top portion of a narrow strip of the weird shape blown up to several times its size.

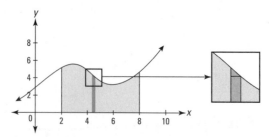

Figure 1-7: For the umpteenth time: When you zoom in, curves become straight.

When you zoom in, the curve becomes practically straight, and the further you zoom in, the straighter it gets — with integration, you actually zoom in *infinitely* close, sort of. You end up with the shape on the right in Figure 1-7, an ordinary trapezoid — or a triangle sitting on top of a rectangle. Because you can compute the areas of rectangles, triangles, and trapezoids with ordinary geometry, you can get the area of this and all the other thin strips and then add up all these areas to get the total area. That's integration.

Why Calculus Works

The mathematics of calculus works because curves are *locally straight*; in other words, they're straight at the microscopic level. The earth is round, but to us it looks flat because we're sort of at the microscopic level when compared to the size of the earth. Calculus works because when you zoom in and curves become straight, you can use regular algebra and geometry with them. This zooming-in process is achieved through the mathematics of limits.

Limits: Math microscopes

The mathematics of *limits* is the microscope that zooms in on a curve. Say you want the exact slope or steepness of the parabola $y = x^2$ at the point (1, 1). See Figure 1-8.

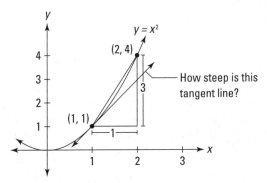

Figure 1-8: The parabola $y = x^2$ with a tangent line at (1, 1).

With the slope formula from algebra, you can figure the slope of the line between (1, 1) and (2, 4) — you go over 1 and up 3, so the slope is 3/1, or 3. But you can see in Figure 1-8 that this line is steeper than the tangent line at (1, 1) that shows the parabola's steepness at that specific point. The limit process sort of lets you slide the point that starts at (2, 4) down toward (1, 1) till it's a thousandth of an inch away, then a millionth, then a billionth, and so on down to the microscopic level. If you do the math, the slopes between (1, 1) and your moving point would look something like 2.001, 2.000001, 2.000000001, and so on. And with the almost magical mathematics of limits, you can conclude that the slope at (1, 1) is precisely 2, even though the sliding point never reaches (1, 1). (If it did, you'd only have one point left and you need two separate points to use the slope formula.) The mathematics of limits is all based on this zooming-in process, and it works, again, because the further you zoom in, the straighter the curve gets.

What happens when you zoom in

Figure 1-9 shows three diagrams of one curve and three things you might like to know about the curve: 1) the exact slope or steepness at point C, 2) the area under the curve between A and B, and 3) the exact length of the curve from A to B. You can't answer these questions with regular math because the regular math formulas for *slope*, *area*, and *length* work for straight lines (and simple curves like circles), but not for weird curves like this one.

The first row of Figure 1-10 shows a magnified detail from the three diagrams of the curve in Figure 1-9. The second row shows further magnification. Each magnification makes the curves straighter and straighter and closer and closer to the diagonal line. This process is continued indefinitely.

Finally, the bottom row of Figure 1-10 shows the result after an "infinite" number of magnifications — sort of. You can think of the lengths 3 and 4 in the bottom row of rectangles as 3 and 4 millionths of an inch, no, make that 3 and 4 billionths of an inch, no, trillionths, no, gazillionths, ...

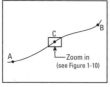
How steep is the curve at C?

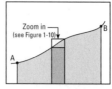
What's the area under the curve between A and B?

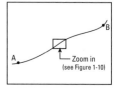
What's the length of the curve from A to B?

Figure 1-9: One curve — three questions.

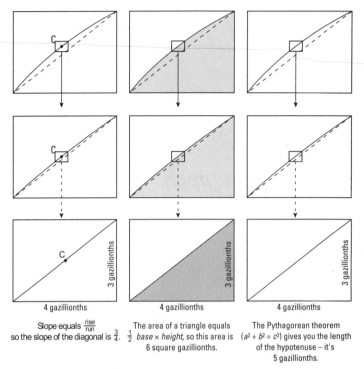

Slope equals $\frac{rise}{run}$ so the slope of the diagonal is $\frac{3}{4}$.

The area of a triangle equals $\frac{1}{2}$ base × height, so this area is 6 square gazillionths.

The Pythagorean theorem ($a^2 + b^2 = c^2$) gives you the length of the hypotenuse – it's 5 gazillionths.

Figure 1-10: Zooming in to the sub, sub, sub . . . subatomic level.

After zooming in "forever," the curve is perfectly straight, and now regular algebra and geometry formulas work.

For the diagram on the left in Figure 1-10, you can now use the regular *slope* formula from algebra to find the slope at point C. It's exactly 3/4 — that's the answer to the first question in Figure 1-9. This is how differentiation works.

For the diagram in the middle, the regular triangle formula from geometry gives an area of 6. So to get the total shaded area in Figure 1-9, you add the area of the thin rectangle under this triangle (the thin strip in Figure 1-9 shows the basic idea), repeat this process for all the other narrow strips, and then just add up all the little areas. This is how regular integration works.

For the diagram on the right, geometry's Pythagorean theorem gives you a length of 5. Then to find the total length of the curve from A to B in Figure 1-9, you do the same thing for the other minute sections of the curve and then add up all the little lengths. This is how you calculate arc length (another integration problem).

Well, there you have it. Calculus uses the limit process to zoom in on a curve till it's straight. After it's straight, the rules of regular-old math apply. Calculus thus gives ordinary algebra and geometry the power to handle complicated problems involving *changing* quantities (which on a graph show up as *curves*). This explains why calculus has so many practical uses, because if there's something you sure can count on — in addition to death and taxes — it's that things are always changing.

Chapter 2

Limits and Continuity

• •

In This Chapter

▶ Taking a look at limits

▶ Evaluating functions with holes

▶ Exploring continuity and discontinuity

• •

*L*imits are fundamental for both differential and integral calculus. The formal definition of a derivative involves a limit as does the definition of a definite integral.

Taking It to the Limit

The *limit* of a function (if it exists) for some *x*-value, *a,* is the height the function gets closer and closer to as *x* gets closer and closer to *a* from the left and the right.

Let me say that another way. A function has a limit for a given *x*-value if the function zeroes in on some height as *x* gets closer and closer to the given value from the left and the right. It's easier to understand limits through examples than through this sort of mumbo jumbo, so take a look at some.

Three functions with one limit

Consider the function $f(x) = 3x - 1$ in Figure 2-1. When we say that the limit of $f(x)$ as x approaches 2 is 5, written as $\lim_{x \to 2} f(x) = 5$, we mean that as x gets closer and closer to 2 from the left and the right, $f(x)$ gets closer and closer to a height of 5. (As far as I know, the number 2 in this example doesn't have a formal name, but I call it the *x-number*.) Look at Table 2-1.

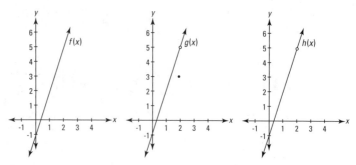

Figure 2-1: The graphs of $f(x)$, $g(x)$, and $h(x)$

Table 2-1   Input and Output Values of $f(x) = 3x-1$ as x Approaches 2

	x approaches 2 from the left					x approaches 2 from the right				
x	1	1.5	1.9	1.99	1.999	2.001	2.01	2.1	2.5	3
$f(x)$	2	3.5	4.7	4.97	4.997	5.003	5.03	5.3	6.5	8
	y approaches 5					y approaches 5				

In Table 2-1, y is approaching 5 as x approaches 2 from both the left and the right, and thus the limit is 5. But why all the fuss with the numbers in the table? Why not just plug the number 2 into $3x - 1$ and obtain the answer of 5? In fact, if all functions were *continuous* (without gaps) as in Figure 2-1, you *could* just plug in the x-number to get the answer, and limit problems would basically be pointless.

We need limits in calculus because of functions like the ones in Figure 2-1 that have holes. g is identical to f except for the hole at (2, 5) and the point at (2, 3). h is identical to f except for the hole at (2, 5).

Imagine what the table of input and output values would look like for g and h. Can you see that the values would be identical to the values in Table 2-1 for f? For both g and h, as x gets closer and closer to 2 from the left and the right, y gets closer and closer to a height of 5. For all three functions, the limit as x approaches 2 is 5.

This brings us to a critical point: When determining the limit of a function as x approaches, say, 2, the value of the function when $x = 2$ is totally irrelevant. Consider the three functions where $x = 2$: $f(2)$ equals 5, $g(2)$ is 3, and $h(2)$ doesn't exist (is undefined). But, again, those three results are irrelevant and don't affect the answer to the limit problem.

In a limit problem, x gets closer and closer to the x-number but technically *never gets there,* and what happens to the function when x equals the x-number has *no effect* on the answer to the limit problem (though for continuous functions like f the function value equals the limit answer and it can thus be used to compute the limit answer).

One-sided limits

One-sided limits work like regular, two-sided limits except that x approaches the x-number from just the left or just the right. To indicate a one-sided limit, you put a superscript subtraction sign on the x-number when x approaches the x-number from the left or a superscript addition sign when x approaches it from the right:

$$\lim_{x \to 4^-} f(x) \quad \text{or} \quad \lim_{x \to 10^+} f(x)$$

Look at Figure 2-2. The answer to the regular limit problem, $\lim_{x \to 3} p(x)$, is that the limit does not exist because as x approaches 3 from the left and the right, $p(x)$ is not zeroing in on the same height.

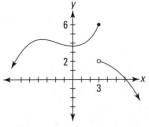

Figure 2-2: $p(x)$: An illustration of two one-sided limits.

However, both one-sided limits do exist. As x approaches 3 from the left, p zeros in on a height of 6, and when x approaches 3 from the right, p zeros in on a height of 2. As with regular limits, the value of $p(3)$ has no effect on the answer to either of these one-sided limit problems. Thus,

$$\lim_{x \to 3^-} p(x) = 6 \quad \text{and} \quad \lim_{x \to 3^+} p(x) = 2$$

Limits and vertical asymptotes

A *rational* function like $f(x) = \dfrac{(x+2)(x-5)}{(x-3)(x+1)}$ has vertical asymptotes at $x = 3$ and $x = -1$. Remember asymptotes? They're imaginary lines that a function gets closer and closer to as it goes up, down, left, or right toward infinity. f is shown in Figure 2-3.

Consider the limit of f as x approaches 3. As x approaches 3 from the left, f goes up to infinity; and as x approaches 3 from the right, f goes down to negative infinity. Sometimes it's informative to indicate this by writing,

$$\lim_{x \to 3^-} f(x) = \infty \quad \text{and} \quad \lim_{x \to 3^+} f(x) = -\infty$$

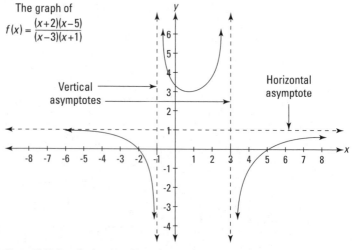

The graph of
$$f(x) = \frac{(x+2)(x-5)}{(x-3)(x+1)}$$

Vertical asymptotes

Horizontal asymptote

Figure 2-3: A typical rational function.

But it's also correct to say that both of the above limits *do not exist* because infinity is not a real number. And if you're asked to determine the regular, two-sided limit, $\lim_{x \to 3} f(x)$, you have to say that it does not exist because the limits from the left and from the right are unequal.

Limits and horizontal asymptotes

Up till now, I've been looking at limits where x approaches a regular, finite number. But x can also approach infinity or negative infinity. Limits at infinity exist when a function has a horizontal asymptote. For example, the function in Figure 2-3 has a horizontal asymptote at $y = 1$, which the function crawls along as it goes toward infinity to the right and negative infinity to the left. (Going left, the function crosses the horizontal asymptote at $x = -7$ and then gradually comes down toward the asymptote.) The limits equal the height of the horizontal asymptote and are written as

$$\lim_{x \to \infty} f(x) = 1 \quad \text{and} \quad \lim_{x \to -\infty} f(x) = 1$$

Instantaneous speed

Say you decide to drop a ball out of your second-story window. Here's the formula that tells you how far the ball has dropped after a given number of seconds (ignoring air resistance):

$$h(t) = 16t^2$$

(where h is the height the ball has fallen, in feet, and t is the amount of time since the ball was dropped, in seconds)

If you plug 1 into t, h is 16; so the ball falls 16 feet during the first second. During the first 2 seconds, it falls a total of $16 \cdot 2^2$, or 64 feet, and so on. Now, what if you wanted to determine the ball's speed exactly 1 second after you dropped it? You can start by whipping out this trusty ol' formula:

Distance = *rate* $\cdot$ *time*, so *Rate* = *distance* / *time*

Using the *rate,* or *speed* formula, you can easily figure out the ball's average speed during the 2nd second of its fall. Because it dropped 16 feet after 1 second and a total of 64 feet after 2 seconds, it fell $64 - 16$, or 48 feet from $t = 1$ second to $t = 2$ seconds. The following formula gives you the average speed:

$$Average\ speed = \frac{total\ distance}{total\ time}$$
$$= \frac{64-16}{2-1}$$
$$= 48\ feet\ per\ second$$

But this isn't the answer you want because the ball falls faster and faster as it drops, and you want to know its speed exactly 1 second after you drop it. The ball speeds up between 1 and 2 seconds, so this *average* speed of 48 *feet per second* during the 2nd second is certain to be faster than the ball's *instantaneous* speed at the end of the 1st second. For a better approximation, calculate the average speed between $t = 1$ second and $t = 1.5$ seconds. After 1.5 seconds, the ball has fallen $16 \cdot 1.5^2$, or 36 feet, so from $t = 1$ to $t = 1.5$, it falls $36 - 16$, or 20 feet. Its average speed is thus

$$Average\ speed = \frac{36-16}{1.5-1}$$
$$= 40\ feet\ per\ second$$

If you continue this process for elapsed times of a quarter of a second, a tenth, a hundredth, and so on, you arrive at average speeds shown in Table 2-2.

Table 2-2   **Average Speeds from 1 Second to t Seconds**

t seconds	2	$1\frac{1}{2}$	$1\frac{1}{4}$	$1\frac{1}{10}$	$1\frac{1}{100}$	$1\frac{1}{1,000}$	$1\frac{1}{10,000}$
Ave. speed from 1 sec. to t sec.	48	40	36	33.6	32.16	32.016	32.0016

As t gets closer and closer to 1 second, the average speeds appear to get closer and closer to 32 *feet per second.*

Here's the formula I used to generate the numbers in Table 2-2. It gives you the average speed between 1 second and t seconds. Figure 2-4 shows the graph of this function.

$$\text{Average speed} = \frac{16t^2 - 16 \cdot 1^2}{t-1}$$

$$= \frac{16(t^2-1)}{t-1}$$

$$= \frac{16(t-1)(t+1)}{t-1}$$

$$= 16t + 16t \quad (t \neq 1)$$

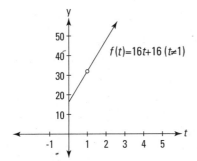

Figure 2-4: The average speed function.

This graph is identical to the graph of the line $y = 16t + 16$ except for the hole at (1, 32). There's a hole there because if you plug 1 into t in the average speed function, you get

$$\text{Average speed} = \frac{16(1^2-1)}{1-1} = \frac{0}{0}$$

which is undefined. And why did you get $\frac{0}{0}$? Because you're trying to determine an average speed — which equals *total distance* divided by *elapsed time* — from $t = 1$ to $t = 1$. But from $t = 1$ to $t = 1$ is, of course, *no* time, and "during" this point in time, the ball doesn't travel any distance, so you get $\frac{\text{zero feet}}{\text{zero seconds}}$ as the average speed from $t = 1$ to $t = 1$.

Obviously, there's a problem here. Hold on to your hat — you've arrived at one of the big "Ah ha!" moments in the development of differential calculus.

Instantaneous speed is defined as the limit of the average speed as the elapsed time approaches zero.

The fact that the elapsed time never gets to zero doesn't affect the precision of the answer to this limit problem — the answer is exactly 32 *feet per second,* the height of the hole in Figure 2-4. Remarkably, limits let you calculate precise, instantaneous speed at a *single* point in time by taking the limit of a function that's based on an *elapsed* time, a period between *two* points of time.

Limits and Continuity

A *continuous* function is simply a function with no gaps — a function that you can draw without taking your pencil off the paper. Consider the four functions in Figure 2-5.

Whether or not a function is continuous is almost always obvious. The first two functions in Figure 2-5 — f and g — have no gaps, so they're continuous. The next two — p and q — have gaps at $x = 3$, so they're not continuous. That's all there is to it. Well, not quite. The two functions with gaps are not continuous everywhere, but because you can draw sections of them without taking your pencil off the paper, you can say that parts of them are continuous. And sometimes a function is continuous everywhere it's defined. Such a function is described as being *continuous over its entire domain,* which means that its gap or gaps occur at x-values where the function is undefined. The function p is continuous over its entire domain; q, on the other hand, is not continuous over its entire domain because it's not continuous at $x = 3$, which is in the function's domain.

Continuity and limits usually go hand in hand. Look at $x = 3$ on the four functions in Figure 2-5. Consider whether each function is continuous there and whether a limit exists at that x-value. The first two, f and g, have no gaps at $x = 3$, so they're continuous there. Both functions also have limits at $x = 3$, and in both cases, the limit equals the height of the function at $x = 3$, because as x gets closer and closer to 3 from the left and the right, y gets closer and closer to $f(3)$ and $g(3)$, respectively.

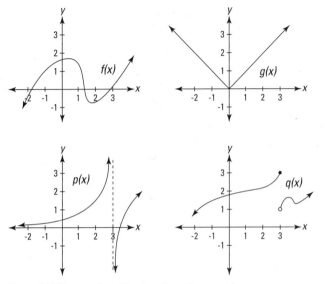

Figure 2-5: The graphs of the functions *f*, *g*, *p*, and *q*.

Functions *p* and *q*, on the other hand, are not continuous at $x = 3$ (or you can say that they're *discontinuous* there), and neither has a regular, two-sided limit at $x = 3$. For both functions, the gaps at $x = 3$ not only break the continuity, they also cause there to be no limits there because, as you move toward $x = 3$ from the left and the right, you do not zero in on some single *y*-value.

So continuity at an *x*-value means there's a limit for that *x*-value, and discontinuity at an *x*-value means there's no limit there . . . except when . . . (read on).

The hole exception

The hole exception is the only exception to the rule that continuity and limits go hand in hand, but it's a *huge* exception. And, I have to admit, it's a bit odd for me to say that continuity and limits *usually* go hand in hand and to talk about this *exception* because the exception is the whole point. When you come right down to it, the exception is more important than the rule. Consider the two functions in Figure 2-6.

These functions have gaps at $x = 2$ and are obviously *not* continuous there, but they *do* have limits as x approaches 2. In each case, the limit equals the height of the hole.

An infinitesimal hole in a function is the only place a function can have a limit where it is not continuous.

So both functions in Figure 2-6 have the same limit as x approaches 2; the limit is 4, and the facts that $r(2) = 1$ and that $s(2)$ is undefined are irrelevant. For both functions, as x zeroes in on 2 from either side, the height of the function zeroes in on the height of the hole — that's the limit.

The limit at a hole is the height of the hole.

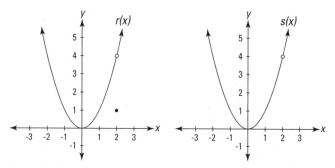

Figure 2-6: The graphs of the functions *r* and *s*.

In the earlier "Instantaneous speed" section, I tried to calculate the average speed during zero elapsed time and got $\frac{zero\ distance}{zero\ time}$. Because $\frac{0}{0}$ is undefined, the result was a hole in the function. Function holes often come about from the impossibility of dividing zero by zero. It's these functions where the limit process is critical, and such functions are at the heart of the meaning of a derivative, and derivatives are at the heart of differential calculus.

A derivative always involves the undefined fraction $\frac{0}{0}$ and always involves the limit of a function with a hole. (All the limits in Chapter 4 — where the derivative is formally defined — are limits of functions with holes.)

Chapter 3

Evaluating Limits

• •

• •

Chapter 2 introduces the concept of a limit. This chapter gets down to the nitty-gritty and presents several techniques for calculating the answers to limit problems.

Easy Limits

A few limit problems are *very* easy.

Limits to memorize

You should memorize the following limits. If you fail to memorize the last three, you'll waste *a lot* of time trying to figure them out.

✔ $\lim\limits_{x \to a} c = c$

($y = c$ is a horizontal line, so the limit — which is the function height — must equal c regardless of the x-number.)

✔ $\lim\limits_{x \to 0^+} \dfrac{1}{x} = \infty$

✔ $\lim\limits_{x \to 0^-} \dfrac{1}{x} = -\infty$

✔ $\lim\limits_{x \to \infty} \dfrac{1}{x} = 0$

✔ $\lim\limits_{x \to -\infty} \dfrac{1}{x} = 0$

> ✔ $\lim\limits_{x\to 0}\dfrac{\sin x}{x}=\lim\limits_{x\to 0}\dfrac{x}{\sin x}=1$
>
> ✔ $\lim\limits_{x\to 0}\dfrac{\cos x-1}{x}=0$
>
> ✔ $\lim\limits_{x\to\infty}\left(1+\dfrac{1}{x}\right)^{x}=e\approx 2.718$

Plug-and-chug limits

Plug-and-chug problems make up the second category of easy limits. Just plug the x-number into the limit function, and if the computation results in a number, that's your answer. For example,

$$\lim_{x\to 3}\left(x^2-10\right)=-1$$

This method works for limits involving continuous functions and functions that are continuous over their entire domains. These are *well-duh* limit problems, and, to be perfectly frank, there's really no point to them. The limit is simply the function value.

The plug-and-chug method works for any type of function *unless* there's a discontinuity at the x-number you plug in. In that case, the number you get after plugging in is *not* the limit. And if you plug the x-number into a limit like $\lim\limits_{x\to 5}\dfrac{10}{x-5}$ and you get any number (other than zero) divided by zero — like $\dfrac{10}{0}$ — then you know that the limit does not exist.

"Real" Limit Problems

If you plug the x-number into the limit expression and the result is undefined (usually $\dfrac{0}{0}$), you've got a "real" limit problem — and some work to do. In this chapter, you learn algebraic techniques for solving these "real" limit problems.

You use two main algebraic techniques for "real" limit problems: factoring and conjugate multiplication. I lump other algebra techniques in the section "Miscellaneous algebra." All algebraic methods involve the same basic idea. When substitution doesn't work in the original function — usually because of a hole in the function — you can use algebra to manipulate the function until substitution does work (it works because your manipulation plugs the hole).

Factoring

Here's an example. Evaluate $\lim\limits_{x \to 5} \dfrac{x^2 - 25}{x - 5}$.

1. **Try plugging 5 into x — you should always try substitution first.**

   You get $\dfrac{0}{0}$ — no good, on to plan B.

2. **$x^2 - 25$ can be factored, so do it.**

   $$\lim_{x \to 5} \frac{x^2 - 25}{x - 5}$$
   $$= \lim_{x \to 5} \frac{(x-5)(x+5)}{x-5}$$

3. **Cancel the $(x - 5)$ from the numerator and denominator.**

   $$= \lim_{x \to 5} (x + 5)$$

4. **Now substitution will work.**

   $$= 5 + 5 = 10$$

The function you got after canceling the $(x - 5)$, namely $(x + 5)$, is identical to the original function, $\dfrac{x^2 - 25}{x - 5}$, except that the hole in the original function at $(5, 10)$ has been plugged. Note that the limit as x approaches 5 is 10, which is the height of the hole at $(5, 10)$.

Conjugate multiplication

Try this method for fraction functions that contain square roots. Conjugate multiplication *rationalizes* the numerator or denominator of a fraction, which means getting rid of square roots. Try this one: Evaluate $\lim\limits_{x \to 4} \dfrac{\sqrt{x} - 2}{x - 4}$.

1. **Try substitution.**

   Plug in 4: that gives you $\dfrac{0}{0}$ — time for plan B.

2. **Multiply the numerator *and* denominator by the conjugate of $\sqrt{x} - 2$, which is $\sqrt{x} + 2$.**

   Now do the rationalizing.

   $$\lim_{x \to 4} \frac{\sqrt{x} - 2}{x - 4}$$

   $$= \lim_{x \to 4} \frac{\left(\sqrt{x} - 2\right)}{\left(x - 4\right)} \cdot \frac{\left(\sqrt{x} + 2\right)}{\left(\sqrt{x} + 2\right)}$$

   $$= \lim_{x \to 4} \frac{\left(\sqrt{x}\right)^2 - 2^2}{\left(x - 4\right)\left(\sqrt{x} + 2\right)}$$

   $$= \lim_{x \to 4} \frac{\left(x - 4\right)}{\left(x - 4\right)\left(\sqrt{x} + 2\right)}$$

3. **Cancel the (x – 4) from the numerator and denominator.**

   $$= \lim_{x \to 4} \frac{1}{\sqrt{x} + 2}$$

4. **Now substitution works.**

   $$= \frac{1}{\sqrt{4} + 2} = \frac{1}{4}$$

As with the factoring example, this rationalizing process plugged the hole in the original function. In this example, 4 is the x-number, $\frac{1}{4}$ is the answer, and the function $\frac{\sqrt{x} - 2}{x - 4}$ has a hole at $\left(4, \frac{1}{4}\right)$.

Miscellaneous algebra

When factoring and conjugate multiplication don't work, try other basic algebra like adding or subtracting fractions, multiplying or dividing fractions, canceling, or some other form of simplification.

Evaluate $\displaystyle\lim_{x \to 0} \frac{\dfrac{1}{x+4} - \dfrac{1}{4}}{x}$.

1. **Try substitution.**

   Plug in 0: That gives you $\frac{0}{0}$ — no good.

2. **Simplify the complex fraction (that's a big fraction that contains little fractions) by multiplying the**

numerator and denominator by the least common denominator of the little fractions, namely 4(x + 4).

Note: Adding or subtracting the little fractions in the numerator or denominator also works in this type of problem, but it's a bit longer than the method here.

$$\lim_{x \to 0} \frac{\left(\dfrac{1}{x+4} - \dfrac{1}{4} \right)}{x} \cdot \frac{4(x+4)}{4(x+4)}$$

$$= \lim_{x \to 0} \frac{4 - (x+4)}{4x(x+4)}$$

$$= \lim_{x \to 0} \frac{-x}{4x(x+4)}$$

$$= \lim_{x \to 0} \frac{-1}{4(x+4)}$$

3. Now substitution works.

$$= \frac{-1}{4(0+4)} = -\frac{1}{16}$$

Limits at Infinity

In the limits in the last section, x approaches a finite number, but there are also limits where x approaches infinity or negative infinity. Consider the function $f(x) = \dfrac{1}{x}$. See Figure 3-1.

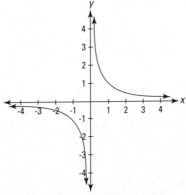

Figure 3-1: The graph of $f(x) = \dfrac{1}{x}$.

You can see on the graph (in the first quadrant) that as x gets bigger and bigger — in other words, as x approaches infinity — the height of the function gets lower and lower but never gets to zero. This is confirmed by considering what happens when you plug bigger and bigger numbers into $\frac{1}{x}$. The outputs get smaller and smaller. This graph thus has a horizontal asymptote of $y = 0$ (the x-axis), and we say that $\lim_{x \to \infty} \frac{1}{x} = 0$. The fact that x never actually reaches infinity and that f never gets to zero has no relevance. When we say that $\lim_{x \to \infty} \frac{1}{x} = 0$, we mean that as x gets bigger and bigger without end, f gets closer and closer to zero. The function f also approaches zero as x approaches negative infinity, written as $\lim_{x \to -\infty} \frac{1}{x} = 0$.

Horizontal asymptotes

Horizontal asymptotes and limits at infinity go hand in hand — you can't have one without the other. For a rational function like $f(x) = \frac{3x - 7}{2x + 8}$, determining the limit at infinity or negative infinity is the same as finding the location of the horizontal asymptote.

Here's what you do. First, note the degree of the numerator (that's the highest power of x in the numerator) and the degree of the denominator. You've got three cases:

- If the degree of the numerator is greater than the degree of the denominator, for example $f(x) = \frac{6x^4 + x^3 - 7}{2x^2 + 8}$, there's no horizontal asymptote and the limit of the function as x approaches infinity (or negative infinity) does not exist.

- If the degree of the denominator is greater than the degree of the numerator, for example $g(x) = \frac{4x^2 - 9}{x^3 + 12}$, the x-axis (the line $y = 0$) is the horizontal asymptote and $\lim_{x \to \infty} g(x) = \lim_{x \to -\infty} g(x) = 0$.

- If the degrees of the numerator and denominator are equal, take the coefficient of the highest power of x in the numerator and divide it by the coefficient of the highest power of x in the denominator. That quotient gives you the answer to the limit problem and the height of

the asymptote. For example, if $h(x) = \dfrac{4x^3 - 10x + 1}{5x^3 + 3x^2 - x}$,
$\lim\limits_{x \to \infty} h(x) = \lim\limits_{x \to -\infty} h(x) = \dfrac{4}{5}$ and h has a horizontal asymptote
at $y = \dfrac{4}{5}$.

WARNING! Substitution doesn't work for the problems in this section. If you try plugging ∞ into x in any of the rational functions in this section, you get $\dfrac{\infty}{\infty}$, but that does *not* equal 1. A result of $\dfrac{\infty}{\infty}$ tells you nothing about the answer to a limit problem.

Solving limits at infinity

Let's try some algebra for the problem $\lim\limits_{x \to \infty}\left(\sqrt{x^2 + x} - x\right)$.

1. Try substitution — always a good idea.

No good. You get $\infty - \infty$, which does *not* equal zero and which tells you nothing (see the related Warning in the previous section). On to plan B.

Because $\left(\sqrt{x^2 + x} - x\right)$ contains a square root, the conjugate multiplication method seems like a natural choice, except that that method is used for fraction functions. Well, just put $\left(\sqrt{x^2 + x} - x\right)$ over the number 1 and, voilà, you've got a fraction: $\dfrac{\sqrt{x^2 + x} - x}{1}$. Now do the conjugate multiplication.

2. Multiply the numerator and denominator by the conjugate of $\left(\sqrt{x^2 + x} - x\right)$ and simplify.

$$\lim_{x \to \infty} \frac{\left(\sqrt{x^2 + x} - x\right)}{1} \cdot \frac{\left(\sqrt{x^2 + x} + x\right)}{\left(\sqrt{x^2 + x} + x\right)}$$

$$= \lim_{x \to \infty} \frac{x^2 + x - x^2}{\sqrt{x^2 + x} + x}$$ (Cancel the x^2s in the numerator. Then factor x out of the denominator;

$$= \lim_{x \to \infty} \frac{x}{x\left(\sqrt{1 + \dfrac{1}{x}} + 1\right)}$$ yes, you heard that right!)

$$= \lim_{x \to \infty} \frac{1}{\sqrt{1 + \dfrac{1}{x}} + 1}$$

3. Now substitution does work.

$$= \frac{1}{\sqrt{1 + \frac{1}{\infty}} + 1}$$

$$= \frac{1}{\sqrt{1+0} + 1} = \frac{1}{2}$$

Chapter 4

Differentiation Orientation

● ●

In This Chapter

▶ Discovering the simple algebra behind the calculus

▶ Finding the derivatives of lines and curves

▶ Tackling the tangent line problem and the difference quotient

● ●

*D*ifferential calculus is the mathematics of *change* and of *infinitesimals*. You might say it's the mathematics of infinitesimal changes — changes that occur every gazillionth of a second.

Without differential calculus — if you've got only algebra, geometry, and trigonometry — you're limited to the mathematics of things that either don't change or that change or move at an *unchanging* rate. Remember those problems from algebra? One train leaves the station at 3 p.m. going west at 80 *mph*. Two hours later another train leaves going east at 50 *mph*. : . . You can handle such a problem with algebra because the speeds or rates are unchanging. Our world, however, isn't one of unchanging rates — rates are in constant flux.

Think about putting men on the moon. Apollo 11 took off from a *moving* launch pad (the earth is both rotating on its axis and revolving around the sun). As the Apollo flew higher and higher, the friction caused by the atmosphere and the effect of the earth's gravity were changing not just every second, not just every millionth of a second, but every *infinitesimal* fraction of a second. The spacecraft's weight was also constantly changing as it burned fuel. All of these things influenced the rocket's changing speed. On top of all that, the rocket had to hit a *moving* target, the moon. All of these things were changing, and their rates of change were changing. Say the rocket was going 1000 *mph* one second and 1020 *mph* a second later — during that one second, the rocket's speed literally passed through the *infinite* number of different speeds between 1000 and 1020 *mph*. How can you do the math for these ephemeral things that change every *infinitesimal* part of a second? You can't do it without differentiation.

Differential calculus is used for all sorts of terrestrial things as well. Modern economic theory, for example, relies on differentiation. In economics, everything is in constant flux. Prices go up and down, supply and demand fluctuate, and inflation is constantly changing. These things are constantly changing, and the ways they affect each other are constantly changing. You need calculus for this.

The Derivative: It's Just Slope

Differentiation is the first of the two major ideas in calculus (the other is integration). Differentiation is the process of finding the derivative of a function like $y = x^2$. The *derivative* is just a fancy calculus term for a simple idea you know from algebra: slope. *Slope,* as you know, is the fancy algebra term for steepness.

In *differential* calculus, you study *differentiation,* which is the process of *deriving* — that's finding — *derivatives*. These are big words for a simple idea: finding the steepness or slope of a line or curve. Throw some of these terms around to impress your friends.

Consider Figure 4-1. A steepness of 1/2 means that as the stickman walks one foot to the right, he goes up 1/2 foot; where the steepness is 3, he goes up 3 feet as he walks 1 foot to the right. Where the steepness is zero, he's at the top, going neither up nor down; and where the steepness is negative, he's going down. A steepness of –2, for example, means that he goes *down* 2 feet for every foot he goes to the right.

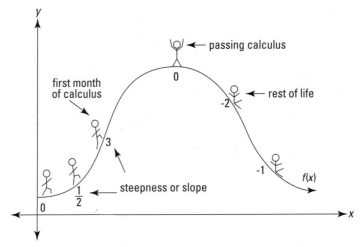

first month
of calculus

passing calculus

0

rest of life
-2

3

-1

f(x)

$\frac{1}{2}$ ← steepness or slope

0

Figure 4-1: Differentiating just means finding the steepness or slope.

The slope of a line

By now you should know that slope is what differentiation is all about. Consider the line $y = 2x + 3$. Plug 1 into x and y equals 5, which gives you the point located at $(1, 5)$; plug 2 into x and y equals 7, giving you the point $(2, 7)$; plug 3 into x and y equals 9, that's the point $(3, 9)$, and so on. Now let's calculate the line's slope (yes, I realize that $y = mx + b$ tells me that the slope is 2; just humor me). Recall that

$$Slope = \frac{rise}{run} = \frac{y_2 - y_1}{x_2 - x_1}$$

The *rise* is the vertical distance between two points, and the *run* is the horizontal distance between the two points. Now, take any two points on the line, say, $(1, 5)$ and $(6, 15)$, and figure the rise and the run. You *rise* up 10 from $(1, 5)$ to $(6, 15)$ because 5 plus 10 is 15. And you *run* across 5 from $(1, 5)$ to $(6, 15)$ because 1 plus 5 is 6. Next, divide to get the slope:

$$Slope = \frac{rise}{run} = \frac{10}{5} = 2$$

Here's how you do the same problem using the slope formula:

$$Slope = \frac{y_2 - y_1}{x_2 - x_1}$$

Plug in the points (1, 5) and (6, 15):

$$Slope = \frac{15 - 5}{6 - 1} = 2$$

The derivative of a line

The preceding section showed you the algebra of slope. Now here's the calculus. The derivative (the slope) of the line, $y = 2x + 3$ is always 2, so you write

$$\frac{dy}{dx} = 2 \quad \text{(Read } dee \; y \; dee \; x \; equals \; 2.)$$

Another common way of writing the same thing is

$$y' = 2 \quad \text{(Read } y \; prime \; equals \; 2.)$$

And you say,

The derivative of the function, $y = 2x + 3$, is 2.

The Derivative: It's Just a Rate

Here's another way to understand the idea of a derivative that's even more fundamental than the concept of slope: A derivative is a *rate*. So why did I start the chapter with *slope*? Because slope is in some respects the easier of the two concepts, and slope is the idea you return to again and again in this book and any other calculus textbook as you look at the graphs of dozens of functions. But before you've got a slope, you've got a rate. A slope is, in a sense, a picture of a rate; the rate comes first, the picture of it comes second. Just like you can have a function before you see its graph, you can have a rate before you see it as a slope.

Calculus on the playground

Imagine Laurel and Hardy on a teeter-totter — see Figure 4-2.

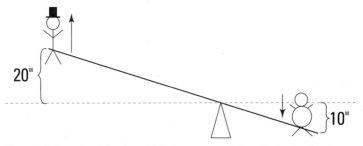

Figure 4-2: Laurel and Hardy — blithely unaware of the calculus implications.

Assuming Hardy weighs twice as much as Laurel, Hardy has to sit twice as close to the center as Laurel for them to balance. And for every inch that Hardy goes down, Laurel goes up two. So Laurel moves twice as much as Hardy. Voilà, you've got a derivative!

A *derivative* is simply a measure of how much one thing changes compared to another — and that's a *rate*.

Laurel moves twice as much as Hardy, so you write

$$dL = 2dH$$

Loosely speaking, dL can be thought of as the change in Laurel's position, and dH as the change in Hardy's position. You can see that if Hardy goes down 10 inches, then dH is 10, and because dL equals 2 times dH, dL is 20 — so Laurel goes up 20 inches. Dividing both sides of this equation by dH gives you

$$\frac{dL}{dH} = 2$$

And that's the derivative of Laurel with respect to Hardy. (It's read as, "dee L, dee H," or as, "the derivative of L with respect to H.") The fact that $\frac{dL}{dH} = 2$ simply means Laurel is moving 2 times as much as Hardy. Laurel's rate of movement is 2 *inches per inch* of Hardy's movement.

Now let's look at it from Hardy's point of view. Hardy moves half as much as Laurel, so you can also write

$$dH = \frac{1}{2}dL$$

Dividing by dL gives you

$$\frac{dH}{dL} = \frac{1}{2}$$

This is the derivative of Hardy with respect to Laurel, and it means Hardy moves $\frac{1}{2}$ inch for every inch Laurel moves. Thus, Hardy's rate is $\frac{1}{2}$ *inch per inch* of Laurel's movement. By the way, you can also get this derivative by taking $\frac{dL}{dH} = 2$, which is the same as $\frac{dL}{dH} = \frac{2}{1}$, and flipping it upside down to get $\frac{dH}{dL} = \frac{1}{2}$.

These rates of 2 *inches per inch* and $\frac{1}{2}$ *inch per inch* may seem odd because we often think of rates as referring to something per unit of time, like *miles per hour*. But a rate can be *anything per anything*. So, whenever you've got a *this per that*, you've got a rate; and if you've got a rate, you've got a derivative.

There's no end to the different rates you might see: *miles per gallon* (for gas mileage), *gallons per minute* (for water draining out of a pool), *output per employee* (for a factory's productivity), and so on. Rates can be constant or changing. In either case, every rate is a derivative, and every derivative is a rate.

The rate-slope connection

Rates and slopes have a simple connection. The previous rate examples can be graphed on an *x-y* coordinate system, where each rate appears as a slope. Laurel moving twice as much as Hardy can be represented by the following equation:

$$L = 2H$$

Figure 4-3 shows the graph of this function.

The inches on the *H-axis* indicate how far Hardy has moved up or down from the teeter-totter's starting position; the inches on the *L-axis* show how far Laurel has moved up or down. The line goes up 2 inches for each inch it goes to the right, and its slope is thus $\frac{2}{1}$, or 2. This is the visual depiction of $\frac{dL}{dH} = 2$, and it shows that Laurel's position changes 2 times as much as Hardy's.

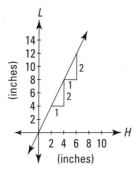

Figure 4-3: The graph of $L = 2H$.

One last comment. You know that $slope = \frac{rise}{run}$. Well, you can think of dL as the rise and dH as the run. That ties everything together.

$$slope = \frac{rise}{run} = \frac{dL}{dH} = rate$$

Don't forget, a derivative is a slope, and a derivative is a rate.

The Derivative of a Curve

The previous sections in this chapter involve *linear* functions — straight lines with *unchanging* slopes. But if all functions and graphs were lines with unchanging slopes, there'd be no need for calculus. The derivative of the Laurel and Hardy function graphed previously is 2, but you don't need calculus to determine the slope of a line. Calculus is the mathematics of change, so now is a good time to move on to *parabolas,* curves with *changing* slopes. Figure 4-4 is the graph of the parabola, $y = \frac{1}{4}x^2$.

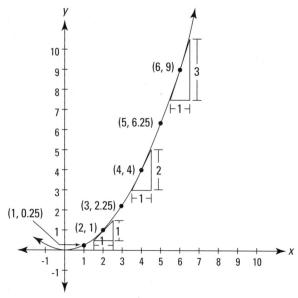

Figure 4-4: The graph of $y = \frac{1}{4}x^2$.

Notice how the parabola gets steeper and steeper as you go to the right. You can see from the graph that at the point (2, 1), the slope is 1; at (4, 4), the slope is 2; at (6, 9), the slope is 3, and so on. It turns out that the derivative of this function equals $\frac{1}{2}x$ (I show you how I got that in a minute). To find the slope of the curve at any point, you just plug the x-coordinate of the point into the derivative, $\frac{1}{2}x$, and you've got the slope. For instance, if you want the slope at the point (3, 2.25), plug 3 into the x, and the slope is $\frac{1}{2}$ times 3, or 1.5.

Here's the calculus. You write

$$\frac{dy}{dx} = \frac{1}{2}x \quad \text{or} \quad y' = \frac{1}{2}x$$

And you say,

The derivative of the function $y = \frac{1}{4}x^2$ is $\frac{1}{2}x$.

I promised to tell you how to *derive* this derivative of $y = \frac{1}{4}x^2$:

1. **Take the power and put it in front of the coefficient.**

$$2 \cdot \frac{1}{4} x^{②}$$

2. **Multiply.**

2 times $\frac{1}{4}$ is $\frac{1}{2}$ so that gives you $\frac{1}{2}x^2$

3. **Reduce the power by 1.**

In this example the 2 becomes a 1. So the derivative is

$\frac{1}{2}x^1$ or just $\frac{1}{2}x$

The Difference Quotient

Sound the trumpets! You come now to what is perhaps the cornerstone of differential calculus: the difference quotient, the bridge between limits and the derivative. (But you're going to have to be patient here, because it's going to take me a few pages to explain the logic behind the difference quotient before I can show you what it is.) Okay, so here goes. You know that a derivative is just a slope. You know all about the slope of a line, and in Figure 4-4, I gave you the slope of the parabola at several points and showed you the short-cut for finding the derivative — but I left out the important math in the middle. That math involves limits and the difference quotient, and takes us to the threshold of calculus. Hold on to your hat.

To compute a slope, you need two points to plug into the slope formula. For a line, this is easy. You just pick any two points on the line and plug them in. But say you want the slope of the parabola in Figure 4-5 at the point (2, 4).

You can see the line drawn tangent to the curve at (2, 4), and because the slope of the tangent line is the same as the slope of the parabola at (2, 4), all you need is the slope of the tangent line. But you don't know the equation of the tangent line, so you can't get the second point — in addition to (2, 4) — that you need for the slope formula. Here's how the inventors of calculus got around this roadblock. Figure 4-5 also shows a *secant* line (a line intersecting a curve at two points) intersecting the parabola at (2, 4) and at (10, 100).

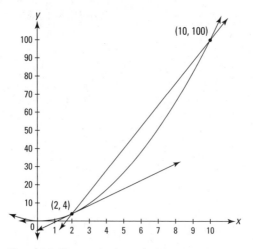

Figure 4-5: The graph of $y = x^2$ with a tangent line and a secant line.

The slope of this secant line is given by the slope formula:

$$Slope = \frac{y_2 - y_1}{x_2 - x_1} = \frac{100 - 4}{10 - 2} = \frac{96}{8} = 12$$

This secant line is quite a bit steeper than the tangent line, so the slope of the secant, 12, is higher than the slope you're looking for.

Now add one more point at (6, 36) and draw another secant using that point and (2, 4) again. See Figure 4-6.

Calculate the slope of this second secant:

$$Slope = \frac{36 - 4}{6 - 2} = \frac{32}{4} = 8$$

You can see that this secant line is a better approximation of the tangent line than the first secant. Now, imagine what would happen if you grabbed the point at (6, 36) and slid it down the parabola toward (2, 4), dragging the secant line along with it. Can you see that as the point gets closer and closer to (2, 4), the secant line gets closer and closer to the tangent line, and that the slope of this secant thus gets closer and closer to the slope of the tangent?

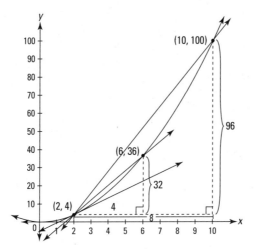

Figure 4-6: The graph of $y = x^2$ with a tangent line and two secant lines.

So, you can get the slope of the tangent if you take the *limit* of the slope of this moving secant. Let's give the moving point the coordinates (x_2, y_2). As this point (x_2, y_2) slides closer and closer to (x_1, y_1), namely (2, 4), the *run* — that's $(x_2 - x_1)$ — gets closer and closer to zero. So here's the limit you need:

$$Slope_{of\ tangent} = \lim_{\substack{as\ point\ slides \\ toward\ (2,\ 4)}} \left(slope_{of\ moving\ secant} \right)$$

$$= \lim_{run \to 0} \frac{rise}{run}$$

$$= \lim_{run \to 0} \frac{y_2 - y_1}{x_2 - x_1}$$

$$= \lim_{run \to 0} \frac{y_2 - 4}{x_2 - 2}$$

Watch what happens to this limit when you plug in three more points on the parabola that are closer and closer to (2, 4):

When the point (x_2, y_2) slides to (2.1, 4.41), the slope is 4.1.

When the point slides to (2.01, 4.0401), the slope is 4.01.

When the point slides to (2.001, 4.004001), the slope is 4.001.

Sure looks like the slope is headed toward 4.

As with all limit problems, the variable in this problem, the *run*, *approaches* but never actually gets to zero. If it got to zero — which would happen if you slid the point you grabbed along the parabola until it was actually on top of (2, 4) — you'd have a slope of $\frac{0}{0}$, which is undefined. But, of course, that's precisely the slope you want — the slope of the line when the point *does* land on top of (2, 4). Herein lies the beauty of the limit process. With this limit, you get the *exact* slope of the *tangent* line even though the limit function, $\frac{y_2 - 4}{x_2 - 2}$, generates slopes of *secant* lines.

Here again is the equation for the slope of the tangent line:

$$Slope_{at\,(2,\,4)} = \lim_{run \to 0} \frac{y_2 - 4}{x_2 - 2}$$

And the slope of the tangent line is — yes — the derivative.

The *derivative* of a function $f(x)$ at some number $x = c$, written $f'(c)$, is the slope of the tangent line to f drawn at c.

The slope fraction $\frac{y_2 - 4}{x_2 - 2}$ is expressed with algebra terminology. Now let's rewrite it to give it that highfalutin calculus look. But first, finally, the definition you've been waiting for:

Definition of the Difference Quotient: There's a calculus term for the general slope fraction, $\frac{rise}{run}$ or $\frac{y_2 - y_1}{x_2 - x_1}$, when you write it in the fancy calculus way. A fraction is a *quotient*, right? And both $y_2 - y_1$ and $x_2 - x_1$ are *differences*, right? So, voilà, it's called the *difference quotient.* Here it is:

$$\frac{f(x+h) - f(x)}{h}$$

In the next two pages I show you how $\frac{y_2 - y_1}{x_2 - x_1}$ morphs into the difference quotient.

Okay, let's lay out this morphing process. First, the *run*, $x_2 - x_1$ (in this example, $x_2 - 2$), is called — don't ask me why — h. Next, because $x_1 = 2$ and the *run* equals h, x_2 equals $2 + h$. You then write y_1 as $f(2)$ and y_2 as $f(2+h)$. Making all the substitutions gives you the derivative of x^2 at $x = 2$:

$$f'(2) = \lim_{run \to 0} \frac{rise}{run}$$

$$= \lim_{x_2 \to 2} \frac{y_2 - 4}{x_2 - 2}$$

$$= \lim_{h \to 0} \frac{f(2+h) - f(2)}{(2+h) - 2}$$

$$= \lim_{h \to 0} \frac{f(2+h) - f(2)}{h}$$

$\displaystyle\lim_{h \to 0} \frac{f(2+h) - f(2)}{h}$ is simply the shrinking $\frac{rise}{run}$ stair step you can see in Figure 4-6 as the point slides down the parabola toward (2, 4).

Figure 4-7 is like Figure 4-6 except that instead of exact points like (6, 36) and (10, 100), the sliding point has the general coordinates of $\big(2+h, f(2+h)\big)$ and the rise and the run are expressed in terms of h.

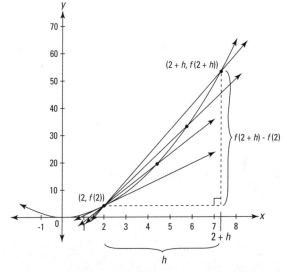

Figure 4-7: Graph of $y = x^2$ showing how a limit produces the slope of the tangent line at (2, 4).

Doing the math gives you the slope of the tangent line at (2, 4):

$$f'(2) = \lim_{h \to 0} \frac{f(2+h) - f(2)}{h}$$

$$= \lim_{h \to 0} \frac{(2+h)^2 - (2)^2}{h}$$

$$= \lim_{h \to 0} \frac{(4 + 4h + h^2) - 4}{h}$$

$$= \lim_{h \to 0} \frac{4h + h^2}{h}$$

$$= \lim_{h \to 0} \frac{h(4+h)}{h}$$

$$= \lim_{h \to 0} (4+h)$$

$$= 4 + 0$$

$$= 4$$

So the slope is 4. (By the way, it's just a coincidence that the slope at (2, 4) happens to be the same as the y-coordinate of the point.)

Definition of the Derivative: If you replace the point $(2, f(2))$ in the above limit equation with the general point $(x, f(x))$, you get the general definition of the derivative as a function of x:

$$f'(x) = \lim_{h \to 0} \frac{f(x+h) - f(x)}{h}$$

So at last you see that the derivative is defined as the limit of the difference quotient.

Now work out this limit and get the derivative for the parabola $f(x) = x^2$.

$$f'(x) = \lim_{h \to 0} \frac{f(x+h) - f(x)}{h}$$

$$= \lim_{h \to 0} \frac{(x+h)^2 - (x)^2}{h}$$

$$= \lim_{h \to 0} \frac{(x^2 + 2xh + h^2) - x^2}{h}$$

$$= \lim_{h \to 0} \frac{2xh + h^2}{h}$$

$$= \lim_{h \to 0} \frac{h(2x + h)}{h}$$

$$= \lim_{h \to 0} (2x + h)$$

$$= 2x + 0$$

$$= 2x$$

Thus for this parabola, the derivative — that's the slope of the tangent line — equals $2x$. Plug any number into x, and you get the slope of the parabola at that x-value! Try it.

Average and Instantaneous Rate

Returning once again to the connection between slopes and rates, a slope is just the visual depiction of a rate: The slope, $\frac{rise}{run}$, just tells you the rate at which y changes compared to x. If, for example, y is the number of miles and x is the number of hours, you get the familiar rate of *miles per hour*.

Each secant line in Figure 4-6 has a slope given by the formula $\frac{y_2 - y_1}{x_2 - x_1}$. That slope is the *average* rate over the interval from x_1 to x_2. If y is in miles and x is in hours, you get the *average* speed in *miles per hour* during the time interval from x_1 to x_2.

When you take the limit and get the slope of the tangent line, you get the *instantaneous* rate at the point (x_1, y_1). Again, if y is in miles and x is in hours, you get the *instantaneous* speed at the point in time, x_1. Because the slope of the tangent line is the derivative, this gives us another definition of the derivative.

The *derivative* of a function $f(x)$ at some x-value is the *instantaneous* rate of change of f with respect to x at that value.

Three Cases Where the Derivative Does Not Exist

By now you certainly know that the derivative of a function at a given point is the slope of the tangent line at that point. So, if you can't draw a tangent line, there's no derivative — that happens in two cases. In the third case, there's a tangent line, but its slope and the derivative are undefined:

1. There's no tangent line and thus no derivative at any type of *discontinuity*: infinite, removable, or jump. Continuity is, therefore, a *necessary* condition for differentiability. It's not, however, a *sufficient* condition as the next two cases show. Dig that logician-speak.

2. There's no tangent line and thus no derivative at a sharp *corner* on a function (or at a *cusp*, a really pointy, sharp turn). See function *f* in Figure 4-8.

3. Where a function has a *vertical inflection point* (see function *g*), the slope is undefined and thus the derivative fails to exist.

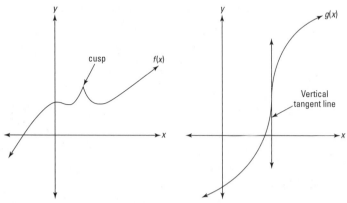

Figure 4-8: Cases 2 and 3 where there's no derivative.

Chapter 5

Differentiation Rules

· ·

· ·

*I*n this chapter I give you shortcut techniques for finding derivatives that avoid the difficulties of limits and the difference quotient. Then, after you absorb this somewhat tedious material, Chapters 6 and 7 show you how to use the derivative to solve all sorts of interesting problems.

Basic Differentiation Rules

Calculus can be difficult, but you'd never know it judging by this section. Learning these few basic rules is a snap.

The constant rule

This is simple. $f(x) = 5$ is a horizontal line with a slope of zero, and thus its derivative is also zero. So, for any number c, if $f(x) = c$, then $f'(x) = 0$. Or you can write $\frac{d}{dx}c = 0$. End of story.

The power rule

Say $f(x) = x^5$. To find its derivative, take the power, 5, bring it in front of the x, and reduce the power by 1 (in this example, the power becomes a 4). That gives you $f(x) = 5x^4$. Piece o' cake.

In Chapter 4, I differentiate $y = x^2$ with the difference quotient. It takes eight lines of math to do it. Instead of all that, just use the power rule: Bring the 2 in front, reduce the power by 1, which leaves you with a power of 1 that you can drop (because a power of 1 does nothing). Thus,

$$y = x^2$$
$$y' = 2x$$

Once you know the nifty shortcuts like this one, you'll rarely use the difference quotient. So why bother learning all that difficult difference quotient stuff? Well, the difference quotient is included in every calculus book and course because it gives you a fuller, richer understanding of the calculus and its foundations — think of it as a mathematical character builder. *Or* because math teachers are sadists. You be the judge.

The power rule works for positive, negative, and fraction powers.

If $f(x) = x^{-2}$ then $f'(x) = -2x^{-3}$

If $g(x) = x^{2/3}$ then $g'(x) = \frac{2}{3}x^{-1/3}$

If $h(x) = x$ then $h'(x) = 1$

Make sure you remember how to do the last function. It's the simplest function, yet the easiest problem to miss. You can just memorize that the derivative of x is 1. Or do it like this: Rewrite $h(x) = x$ as $h(x) = x^1$ and then apply the power rule: Bring the 1 in front and reduce the power by 1 to zero, giving you $h'(x) = 1x^0$. Because x^0 equals 1, you've got $h'(x) = 1$.

You can differentiate radical functions by rewriting them as power functions and then using the power rule. For example, if $f(x) = \sqrt[3]{x^2}$, rewrite it as $f(x) = x^{2/3}$ and use the power rule. You can also use the power rule to differentiate functions like $f(x) = \frac{1}{x^3}$. Rewrite this as $f(x) = x^{-3}$ and use the power rule.

The constant multiple rule

What if the function you're differentiating begins with a coefficient? Makes no difference. A coefficient has no effect on differentiation. Just ignore it and differentiate according to the appropriate rule. The coefficient stays where it is until the final step when you simplify your answer by multiplying by the coefficient.

Differentiate $y = 4x^3$.

Solution: You know by the power rule that the derivative of x^3 is $3x^2$, so the derivative of $4(x^3)$ is $4(3x^2)$. The 4 just sits there doing nothing. Then, as a final step, you simplify: $4(3x^2)$ equals $12x^2$. So $y' = 12x^2$.

Differentiate $y = 5x$.

Solution: This is a line of the form $y = mx + b$ with $m = 5$, so the slope is 5 and thus the derivative is 5: $y' = 5$. (It's important to think graphically like this sometimes.) But you can also solve it with the power rule. $\dfrac{d}{dx}x^1 = 1x^0 = 1$; so $\dfrac{d}{dx}5(x^1) = 5(1) = 5$.

In a nutshell, the constant multiple rule takes a function like $f(x) = 10(stuff)$, differentiates the *stuff* — that's *stuff'* — while the 10 just stays put. So, if $g(x) = 15(stuff)$, then $g'(x) = 15(stuff')$.

Don't forget that π (~3.14) and e (~2.72) are numbers, not variables, so they behave like ordinary numbers. Constants in problems, like c and k, also behave like ordinary numbers.

Thus, if $y = \pi x$, $y' = \pi$ — this works exactly like differentiating $y = 5x$. And because π^3 is just a number, if $y = \pi^3$ then $y' = 0$ — this works exactly like differentiating $y = 10$. You'll also see problems containing constants like c and k. Be sure to treat them like regular numbers. For example, the derivative of $y = 5x + 2k^3$ (where k is a constant) is 5, not $5 + 6k^2$.

The sum and difference rules

When you want the derivative of a sum of terms, take the derivative of each term separately.

What's $f'(x)$ if $f(x) = x^6 + x^3 + x^2 + x + 10$?

Just use the power rule for each of the first four terms and the constant rule for the final term. Thus, $f'(x) = 6x^5 + 3x^2 + 2x + 1$.

If you've got a difference (that's subtraction) instead of a sum, it makes no difference. You still differentiate each term separately. Thus, if $y = 3x^5 - x^4 - 2x^3 + 6x^2 + 5x$, then $y' = 15x^4 - 4x^3 - 6x^2 + 12x + 5$.

Differentiating trig functions

I have the high honor and distinct privilege of introducing you to the derivatives of the six trig functions.

$$\frac{d}{dx}\sin x = \cos x \qquad \frac{d}{dx}\csc x = -\csc x \cot x$$

$$\frac{d}{dx}\cos x = -\sin x \qquad \frac{d}{dx}\sec x = \sec x \tan x$$

$$\frac{d}{dx}\tan x = \sec^2 x \qquad \frac{d}{dx}\cot x = -\csc^2 x$$

Make sure you memorize the derivatives for sine and cosine — they're a snap. You should learn the other four as well, but if you're afraid that this knowledge will crowd out the date of the Battle of Hastings (1066), you can use the following nifty mnemonic device I made up.

Imagine you're taking a test and can't remember these four derivatives. You lean over to the guy next to you and whisper, "Psst, hey what's the derivative of csc$x$?" Now, take the last three letters of *psst* (sst) — those are the initial letters of sec, sec, tan. Write these three down, and below them write their cofunctions: csc, csc, cot. Put a negative sign on the csc in the middle. Finally, add arrows like in the following diagram.

$$sec \;\rightarrow\; sec \;\leftarrow\; tan$$
$$csc \;\rightarrow\; -csc \;\leftarrow\; cot$$

The *sec* on the left has an arrow pointing to *sec tan* — so the derivative of sec$x$ is sec$x$ tan$x$. The *tan* on the right has an arrow pointing to *sec sec*, so the derivative of tan$x$ is $\sec^2 x$. The bottom row works the same except both derivatives are negative. Believe it or not, this trick is easy to remember.

Exponential and logarithmic functions

If you can't memorize the next rule, hang up your calculator.

$$\frac{d}{dx}e^x = e^x$$

That's right, the derivative of e^x is itself! This is a special function. e^x and its multiples, like $5e^x$, are the only functions that are their own derivatives.

If the base is a number other than e, you have to tweak the derivative by multiplying it by the natural log of the base:

If $y = 2^x$, then $y' = 2^x \ln 2$.

If $y = 10^x$, then $y' = 10^x \ln 10$.

Here's the derivative of the *natural* log:

$$\frac{d}{dx}\ln x = \frac{1}{x}$$

If the log base is a number other than e, you tweak this derivative — as with exponential functions — except you *divide* by the natural log of the base instead of multiplying:

$$\frac{d}{dx}\log_2 x = \frac{\frac{1}{x}}{\ln 2} = \frac{1}{x \ln 2}, \text{ and}$$

$$\frac{d}{dx}\log x = \frac{1}{x \ln 10} \qquad \text{(Recall that } \log_{10} x \text{ is written without the 10.)}$$

Derivative Rules for Experts

These rules, especially the chain rule, can be a bit tough.

The product and quotient rules

The Product Rule (for the *product* (duh) of two functions):

If $y = this \cdot that$,

then $y' = this' \cdot that + this \cdot that'$

So, for $y = x^3 \cdot \sin x$,

$$y' = \left(x^3\right)' \cdot \sin x + x^3 \cdot \left(\sin x\right)'$$
$$= 3x^2 \sin x + x^3 \cos x$$

The Quotient Rule (bet you can guess what this is for):

If $y = \dfrac{top}{bottom}$,

then $y' = \dfrac{top' \cdot bottom - top \cdot bottom'}{bottom^2}$

Most calculus books give this rule in a slightly different form that's harder to remember. And some give a "mnemonic" involving the words *lodeehi* and *hideelo* or *hodeehi* and *hideeho*, which is easy to get mixed up — great, thanks a lot.

Memorize the quotient rule as I've written it. You'll remember what goes in the denominator — no one ever forgets it. The trick is knowing the order of the terms in the numerator. Think of it like this: You're doing a derivative, so the first thing you do is to take a derivative. The natural place to begin is at the top of the fraction. So the quotient rule begins with the derivative of the top. Remember that, and the rest of the numerator is almost automatic.

Here's the derivative of $y = \dfrac{\sin x}{x^4}$:

$$y' = \frac{\left(\sin x\right)' \cdot x^4 - \sin x \cdot \left(x^4\right)'}{\left(x^4\right)^2}$$

$$= \frac{x^4 \cos x - 4x^3 \sin x}{x^8}$$

$$= \frac{x^3 \left(x \cos x - 4 \sin x\right)}{x^8}$$

$$= \frac{x \cos x - 4 \sin x}{x^5}$$

The chain rule

The chain rule is by far the trickiest derivative rule, but it's not really that bad if you carefully focus on a few important points. Let's begin by differentiating $y = \sqrt{4x^3 - 5}$. You use the chain rule here because you've got a *composite* function, that's one function $\left(4x^3 - 5\right)$ inside another function (the square root function).

Here's one way to quickly recognize a composite function. $y = \sqrt{x}$ is *not* a composite function because the *argument* of the square root — the thing you take the square root of — is simply x. Whenever the argument of a function is anything other than a plain old x, you've got a composite function. Be careful to distinguish a composite function from something like $y = \sqrt{x} \cdot \sin x$, which is the *product* of two functions, $\sqrt{x}$ and $\sin x$, each of which *does* have just an x as its argument.

Okay, so you've got this composite function, $y = \sqrt{4x^3 - 5}$. Here's how to differentiate it with the chain rule.

1. **You start with the *outside* function, $\sqrt{}$, and differentiate that, *IGNORING* what's inside. To make sure you ignore the inside, temporarily replace the inside function with the word *stuff*.**

   So you've got $y = \sqrt{stuff}$. Okay, now differentiate $y = \sqrt{stuff}$ the same way you'd differentiate $y = \sqrt{x}$. Because $y = \sqrt{x}$ is the same as $y = x^{1/2}$, the power rule gives you $y' = \frac{1}{2}x^{-1/2}$. So for this problem, you begin with $\frac{1}{2} stuff^{-1/2}$.

2. **Multiply the result from Step 1 by the derivative of the inside function, *stuff'*.**

   $$y' = \frac{1}{2} stuff^{-1/2} \cdot stuff'$$

   Take a good look at this. *All* basic chain rule problems follow this format. You do the derivative rule for the outside function, ignoring the inside *stuff*, then multiply that by the derivative of the *stuff*.

3. **Differentiate the inside *stuff*.**

   The inside *stuff* in this problem is $4x^3 - 5$, and its derivative is $12x^2$ by the power rule.

4. **Now put the real *stuff* and its derivative back where they belong.**

$$y' = \frac{1}{2}\left(4x^3 - 5\right)^{-1/2}\left(12x^2\right)$$

$$= 6x^2\left(4x^3 - 5\right)^{-1/2}$$

Or, if you've got something against negative powers, $y' = \dfrac{6x^2}{\left(4x^3 - 5\right)^{1/2}}$. Or, if you've got something against fraction powers, $y' = \dfrac{6x^2}{\sqrt{4x^3 - 5}}$.

Let's differentiate another composite function: $y = \sin\left(x^2\right)$.

1. **The outside function is the sine function, so you start there, taking the derivative of sine and ignoring the inside *stuff*, x^2. The derivative of sin x is cos x, so for this problem, you begin with**

   $\cos(\text{stuff})$

2. **Multiply the derivative of the outside function by the derivative of the *stuff*.**

   $$y' = \cos(\text{stuff}) \cdot \text{stuff}'$$

3. **The *stuff* in this problem is x^2, so *stuff'* is $2x$. When you plug these terms back in, you get**

   $$y' = \cos\left(x^2\right) \cdot 2x$$

   $$= 2x\cos\left(x^2\right)$$

Sometimes figuring out which function is inside which can be tricky — especially when a function is inside another and then both of them are inside a third (you can have four or more nested functions, but three is probably the most you'll see).

Rewrite the composite function with a set of parentheses around each inside function, and rewrite trig functions like $\sin^2 x$ with the power outside a set of parentheses: $(\sin x)^2$.

For example — this is tough — differentiate $y = \sin^3\left(5x^2 - 4x\right)$. First, rewrite the cubed sine function: $y = \left(\sin\left(5x^2 - 4x\right)\right)^3$. Now it's easy to see the order in which the functions are nested. The innermost function is in the innermost parentheses — that's $5x^2 - 4x$. Next, the sine function is in the next set of parentheses — that's $\sin(\textit{stuff})$. Last, the cubing function is outside everything — that's $\textit{stuff}^3$. (Did you notice that the $\textit{stuff}$ in $\textit{stuff}^3$ is different from the $\textit{stuff}$ in $\sin(\textit{stuff})$? I admit this is quite unmathematical, but don't sweat it. I'm just using the term $\textit{stuff}$ to refer to whatever is inside any function.) Now that you know the order of the functions, you differentiate from *outside in*.

1. **The outermost function is $\textit{stuff}^3$ and its derivative is given by the power rule.**

   $3\textit{stuff}^2$

2. **As with all chain rule problems, you multiply that by $\textit{stuff}'$.**

   $3\textit{stuff}^2 \cdot \textit{stuff}'$

3. **Now put the $\textit{stuff}$, $\sin\left(5x^2 - 4x\right)$, back where it belongs.**

   $$3\left(\sin\left(5x^2 - 4x\right)\right)^2 \cdot \left(\sin\left(5x^2 - 4x\right)\right)'$$

4. **Use the chain rule again.**

   You can't finish this quickly by just taking a simple derivative because you have to differentiate another composite function, $\sin\left(5x^2 - 4x\right)$. Just treat $\sin\left(5x^2 - 4x\right)$ as if it were the original problem and take its derivative. The derivative of $\sin x$ is $\cos x$, so the derivative of $\sin(\textit{stuff})$ begins with $\cos(\textit{stuff})$. Multiply that by $\textit{stuff}'$. Thus, the derivative of $\sin(\textit{stuff})$ is

   $\cos(\textit{stuff}) \cdot \textit{stuff}'$

5. **The $\textit{stuff}$ for this step is $5x^2 - 4x$ and its derivative is $10x - 4$. Plug those things back in.**

   $$\cos\left(5x^2 - 4x\right) \cdot \left(10x - 4\right)$$

6. **Now that you've got the derivative of $\sin\left(5x^2 - 4x\right)$, plug this result into the result from Step 3, giving you the whole enchilada.**

   $$3\left(\sin\left(5x^2 - 4x\right)\right)^2 \cdot \left(\cos\left(5x^2 - 4x\right)\right) \cdot \left(10x - 4\right)$$
   $$= \left(30x - 12\right)\sin^2\left(5x^2 - 4x\right)\cos\left(5x^2 - 4x\right)$$

You can save some time by not switching to the word *stuff* and then switching back. But some people like to use the technique because it forces them to leave the *stuff* alone during each step of a problem. That's the critical point.

Make sure you . . . DON'T TOUCH THE *STUFF*.

Don't change the inside function while differentiating the outside one. Say you want to differentiate $f(x) = \ln\left(x^3\right)$. The argument of this natural logarithm function is x^3. Don't touch it during the first step of the solution, which is to use the natural log rule: $\frac{d}{dx}\ln x = \frac{1}{x}$, which says to put the argument of the natural log function in the denominator under the 1. So, after the first step in differentiating $\ln\left(x^3\right)$, you've got $\frac{1}{x^3}$. Finish by multiplying that by the derivative of x^3, which is $3x^2$. Final answer: $\frac{3}{x}$.

Another way to avoid chain rule mistakes is to remember that you never use more than one derivative rule at a time.

In the previous example, $\ln\left(x^3\right)$, you first use the natural log rule, then, as a *separate step,* you use the power rule to differentiate x^3. At no point in any chain rule problem do you use both rules at the same time and write something like $\frac{1}{3x^2}$.

The Chain Rule (for differentiating a composite function):

If $y = f\left(g(x)\right)$,

then $y' = f'\left(g(x)\right) \cdot g'(x)$.

Or, equivalently,

If $y = f(u)$, and $u = g(x)$,

then $\frac{dy}{dx} = \frac{dy}{du} \cdot \frac{du}{dx}$ (notice how the *du*s cancel).

Finally, differentiate $4x^2 \sin\left(x^3\right)$. This one has a new twist — it involves the chain rule *and* the product rule. How do you begin?

If you're not sure where to begin differentiating a complex expression, imagine plugging a number into x and then evaluating the expression on your calculator one step at a time. Your *last* computation tells you the *first* thing to do.

Say you plug the number 5 into the xs in $4x^2 \sin(x^3)$. You evaluate $4 \cdot 5^2$ — that's 100; then, after getting $5^3 = 125$, you do $\sin(125)$, which is about –0.616. Finally, you multiply 100 by –0.616. Because your *last* computation is *multiplication*, your *first* step in differentiating is to use the *product* rule. (Had your last computation been something like $\sin(125)$, you'd begin with the chain rule.) So for $f(x) = 4x^2 \sin(x^3)$, start with the product rule:

$$f'(x) = \left(4x^2\right)'\left(\sin\left(x^3\right)\right) + \left(4x^2\right)\left(\sin\left(x^3\right)\right)'$$

Now finish by taking the derivative of $4x^2$ with the power rule and the derivative of $\sin(x^3)$ with the chain rule:

$$f'(x) = (8x)\left(\sin\left(x^3\right)\right) + \left(4x^2\right)\left(\cos\left(x^3\right) \cdot 3x^2\right)$$
$$= 8x \sin\left(x^3\right) + 12x^4 \cos\left(x^3\right)$$

Differentiating Implicitly

All the differentiation problems so far in this chapter are functions like $y = x^2 + 5x$ or $y = \sin x$ (and y was sometimes written as $f(x)$, as in $f(x) = x^3 - 4x^2$). In such cases, y is written *explicitly* as a function of x. This means that the equation is solved for y (with y by itself on one side of the equation).

Sometimes, however, you are asked to differentiate an equation that's not solved for y, like $y^5 + 3x^2 = \sin x - 4y^3$. This equation defines y *implicitly* as a function of x, and you can't write it as an explicit function because it can't be solved for y. For this you need *implicit differentiation*. When differentiating implicitly, all the derivative rules work the same with one exception: when you differentiate a term with a y in it, you use the chain rule with a little twist.

Remember using the chain rule to differentiate something like $\sin(x^3)$ with the *stuff* technique? The derivative of sine is cosine, so the derivative of $\sin(stuff)$ is $\cos(stuff) \cdot stuff'$. You finish the problem by finding the derivative of the *stuff*, x^3, which is $3x^2$, and then making the substitutions to give you $\cos(x^3) \cdot 3x^2$. With implicit differentiation, a y works just like the word *stuff*. Thus, because

$$\left(\sin(stuff)\right)' = \cos(stuff) \cdot stuff'$$

$$\left(\sin(y)\right)' = \cos(y) \cdot y'$$

The twist is that unlike the word *stuff*, which is temporarily taking the place of some *known* function of x (x^3 in this example), y is some *unknown* function of x. And because you don't know what y equals, the y and the y' — unlike the *stuff* and the *stuff'* — have to remain in the final answer. The concept is the same, and you treat y just like the *stuff*. It's just that because you don't know what the function is, you can't make the switch back to xs at the end of the problem like you can with a regular chain rule problem. Here goes. Differentiate $y^5 + 3x^2 = \sin x - 4y^3$.

1. **Differentiate each term on both sides of the equation.**

   For the first and fourth terms, use the power rule and the chain rule. For the second term, use the regular power rule. For the third, use the regular sine rule.

   $$5y^4 \cdot y' + 6x = \cos x - 12y^2 \cdot y'$$

2. **Collect all terms containing a y' on the left side of the equation and all other terms on the right.**

   $$5y^4 \cdot y' + 12y^2 \cdot y' = \cos x - 6x$$

3. **Factor out y'.**

   $$y'\left(5y^4 + 12y^2\right) = \cos x - 6x$$

4. **Divide for the final answer.**

   $$y' = \frac{\cos x - 6x}{5y^4 + 12y^2}$$

Chapter 6

Differentiation and the Shape of Curves

*I*f you've read Chapters 4 and 5, you're probably an expert at finding derivatives. Which is a good thing, because in this chapter you use derivatives to understand the shape of functions — where they rise, fall, max out and bottom out, how they curve, and so on.

A Calculus Road Trip

Imagine that you're driving along the function in Figure 6-1 from left to right. Along your drive, there are several points of interest between *a* and *l*. All of them, except for the start and finish points, relate to the steepness of the road — in other words, its slope or derivative. I'm going to throw lots of new terms and definitions at you all at once here, but you shouldn't have too much trouble because these ideas mostly involve commonsense notions like driving up or down an incline, or going over the crest of a hill.

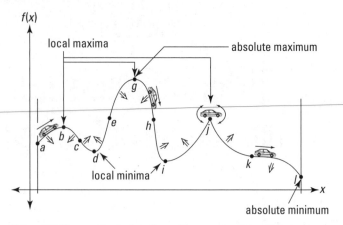

Figure 6-1: The graph of a function with several points of interest.

✔ The function in Figure 6-1 has a derivative of zero at *stationary points* (level points) *b, d, g, i,* and *k.* At *j,* the derivative is *undefined* (a sharp turning point like *j* is a *corner*). These points where the derivative is either zero or undefined are the *critical points* of the function. The *x*-values of these critical points are the function's *critical numbers.*

✔ All *local maxes* and *mins* — peaks and valleys — must occur at critical points. However, not all critical points are local maxes or mins. Point *k,* for instance, is a critical point but neither a max nor a min. Local maximums and minimums — or *maxima* and *minima* — are called, collectively, local *extrema* of the function. A single local max or min is a local *extremum.* Point *g* is the *absolute maximum* on the interval from *a* to *l* because it's the highest point on the road from *a* to *l.* Point *l* is the *absolute minimum.* Note that *g* is also a local max, but *l* is not a local min because endpoints (and points of discontinuity) do not qualify as local extrema.

✔ The function is *increasing* whenever you're going up, where the derivative is *positive*; it's *decreasing* whenever you're going down, where the derivative is *negative.* The function is also decreasing at point *k*, a horizontal inflection point, even though the slope and derivative are zero there. I realize that seems odd, but that's how it works. At all horizontal inflection points, a function is either increasing or decreasing. At local extrema *b, d, g, i,* and *j,* the function is neither increasing nor decreasing.

✔ The function is *concave up* wherever it looks like a cup (rhymes with *up*) or a smile (or where it "holds water") and *concave down* wherever it looks like a frown (rhymes with *down*; where it "spills water"). Wherever a function is concave up, its derivative is increasing; wherever a function is concave down, its derivative is decreasing. *Inflection points c, e, h,* and *k* are where the concavity switches from up to down or vice versa. Inflection points are also the steepest or least steep points in their immediate neighborhoods.

Be careful with function sections that have a negative slope. Point *c* is the steepest point in its neighborhood because it has a bigger negative slope than any other nearby point. But remember, a big negative number is actually a *small* number, so the slope and derivative at *c* are actually the *smallest* of all the points in the neighborhood. From *b* to *c* the derivative of the function is *decreasing* (because it's becoming a bigger negative). From *c* to *d,* it's *increasing* (because it's becoming a smaller negative).

Local Extrema

Now that you know what local extrema are, you need to know how to do the math to find them. Remember, all local extrema occur at critical points of a function — where the derivative is zero or undefined. So, the first step in finding a function's local extrema is to find its critical numbers (the *x*-values of the critical points).

Finding the critical numbers

Find the critical numbers of $f(x) = 3x^5 - 20x^3$. See Figure 6-2.

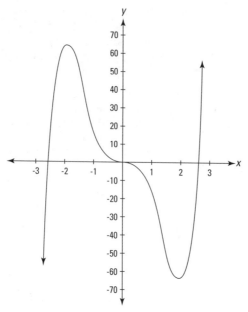

Figure 6-2: The graph of $f(x) = 3x^5 - 20x^3$.

1. **Find the first derivative of f using the power rule.**

$$f(x) = 3x^5 - 20x^3$$
$$f'(x) = 15x^4 - 60x^2$$

2. **Set the derivative equal to zero and solve for x.**

$$15x^4 - 60x^2 = 0$$
$$15x^2(x^2 - 4) = 0$$
$$15x^2(x + 2)(x - 2) = 0$$
$$15x^2 = 0 \quad \text{or} \quad x + 2 = 0 \quad \text{or} \quad x - 2 = 0$$
$$x = 0, \ -2, \quad \text{or} \quad 2$$

These three x-values are critical numbers of f. Additional critical numbers could exist if the first derivative were undefined at some x-values, but because the derivative, $15x^4 - 60x^2$, is defined for all input values, the above solution set, 0, –2, and 2 is the complete list of critical numbers. Because the derivative of f equals zero at these three critical numbers, the curve has horizontal tangents at these numbers.

Now that you've got the list of critical numbers, you need to determine whether peaks or valleys or neither occur at those *x*-values by using either the First Derivative Test or the Second Derivative Test. (Of course, you can see where the peaks and valleys are by just looking at Figure 6-2, but you still have to learn the math.)

The First Derivative Test

The First Derivative Test is based on the Nobel-prize-caliber ideas that as you go over the top of a hill, first you go up and then you go down, and that when you drive into and out of a valley, you go down and then up. This calculus stuff is pretty amazing, isn't it?

Take a number line and put down the critical numbers you found above: 0, –2, and 2. See Figure 6-3.

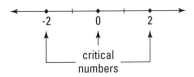

Figure 6-3: The critical numbers of $f(x) = 3x^5 - 20x^3$.

This number line is now divided into four regions: to the left of –2, from –2 to 0, from 0 to 2, and to the right of 2. Pick a value from each region, plug it into the derivative, and note whether your result is positive or negative. Let's use the numbers –3, –1, 1, and 3 to test the regions.

$$f'(x) = 15x^4 - 60x^2$$
$$f'(-3) = 15(-3)^4 - 60(-3)^2 = 15 \cdot 81 - 60 \cdot 9 = 675$$
$$f'(-1) = 15(-1)^4 - 60(-1)^2 = 15 - 60 = -45$$
$$f'(1) = 15(1)^4 - 60(1)^2 = 15 - 60 = -45$$
$$f'(3) = 15(3)^4 - 60(3)^2 = 15 \cdot 81 - 60 \cdot 9 = 675$$

These four results are, respectively, positive, negative, negative, or positive. Now, take your number line, mark each region with the appropriate positive or negative sign, and indicate whether the function is increasing (derivative is positive) or decreasing (derivative is negative). The result is a *sign graph*. See Figure 6-4.

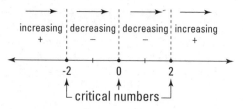

Figure 6-4: The sign graph for $f(x) = 3x^5 - 20x^3$.

Figure 6-4 simply tells you what you already know from the graph of f — that the function goes up until –2, down from –2 to 0, further down from 0 to 2, and up again from 2 on.

The function switches from increasing to decreasing at –2; you go up to –2 and then down. So at –2 you have a hill or a local maximum. Conversely, because the function switches from decreasing to increasing at 2, you have a valley there or a local minimum. And because the signs of the first derivative don't switch at zero, there's neither a min nor a max at that *x*-value (you get an inflection point when this happens).

The last step is to obtain the function values (heights) of these two local extrema by plugging the *x*-values into the original function:

$$f(-2) = 3(-2)^5 - 20(-2)^3 = 64$$
$$f(2) = 3(2)^5 - 20(2)^3 = -64$$

The local max is at $(-2, 64)$ and the local min is at $(2, -64)$.

The Second Derivative Test

The Second Derivative Test is based on two more prize-winning ideas: That at the crest of a hill, a road has a hump shape, so it's curving down or *concave down;* and that at the bottom of a valley, a road is cup-shaped, so it's curving up or *concave up.*

The concavity of a function at a point is given by its second derivative: A *positive* second derivative means the function is concave *up,* a *negative* second derivative means the function is concave *down,* and a second derivative of *zero* is *inconclusive* (the function could be concave up, concave down, or there could be an inflection point there). So, for our function *f*, all you have to do is find its second derivative and then plug in the critical numbers you found (–2, 0, and 2) and note whether your results are positive, negative, or zero. To wit —

$$f(x) = 3x^5 - 20x^3$$
$$f'(x) = 15x^4 - 60x^2 \quad \text{(power rule)}$$
$$f''(x) = 60x^3 - 120x \quad \text{(power rule)}$$

$$f''(-2) = 60(-2)^3 - 120(-2) = -240$$
$$f''(0) = 60(0)^3 - 120(0) = 0$$
$$f''(2) = 60(2)^3 - 120(2) = 240$$

At –2, the second derivative is negative (–240). This tells you that *f* is concave down where *x* equals –2, and therefore that there's a local max at *x* = –2. The second derivative is positive (240) where *x* is 2, so *f* is concave up and thus there's a local min at *x* = 2. Because the second derivative equals zero at *x* = 0, the Second Derivative Test fails — it tells you nothing about the concavity at *x* = 0 or whether there's a local min or max there. When this happens, you have to use the First Derivative Test.

Now go through both derivative tests one more time with another example. Find the local extrema of $g(x) = 2x - 3x^{2/3} + 4$. See Figure 6-5.

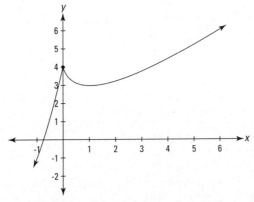

Figure 6-5: The graph of $g(x) = 2x - 3x^{2/3} + 4$.

1. **Find the first derivative of g.**

$$g(x) = 2x - 3x^{2/3} + 4$$
$$g'(x) = 2 - 2x^{-1/3} \quad \text{(power rule)}$$

2. **Set the derivative equal to zero and solve.**

$$2 - 2x^{-1/3} = 0$$
$$-2x^{-1/3} = -2$$
$$x^{-1/3} = 1$$
$$\left(x^{-1/3}\right)^{-3} = 1^{-3}$$
$$x = 1$$

Thus 1 is a critical number.

3. **Determine whether the first derivative is undefined for any x-values.**

$2x^{-1/3}$ equals $\dfrac{2}{\sqrt[3]{x}}$. Now, because the cube root of zero is zero, if you plug in zero to $\dfrac{2}{\sqrt[3]{x}}$, you'd have $\dfrac{2}{0}$, which is undefined. So the derivative, $2 - 2x^{-1/3}$ is undefined at $x = 0$, and thus 0 is another critical number.

4. **Plot the critical numbers on a number line, and then use the First Derivative Test to figure out the sign of each region.**

You can use –1, 0.5, and 2 as test numbers

$$g'(-1) = 4$$
$$g'(0.5) \approx -0.52$$
$$g'(2) \approx 0.41$$

Figure 6-6 shows the sign graph.

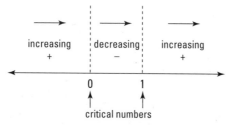

Figure 6-6: The sign graph of $g(x) = 2x - 3x^{2/3} + 4$.

Because the first derivative of g switches from positive to negative at 0, there's a local max there. And because the first derivative switches from negative to positive at 1, there's a local min at $x = 1$.

5. **Plug the critical numbers into g to obtain the function values (the heights) of these two local extrema.**

$$g(x) = 2x - 3x^{2/3} + 4 \qquad g(0) = 4$$
$$g(1) = 3$$

So, there's a local max at (0, 4) and a local min at (1, 3).

You could have used the Second Derivative Test instead of the First in Step 4. You'd need the second derivative of g, which is, as you know, the derivative of its first derivative:

$$g'(x) = 2 - 2x^{-1/3}$$
$$g''(x) = \frac{2}{3}x^{-4/3}$$

Evaluate the second derivative at 1 (the critical number where $g' = 0$).

$$g''(1) = \frac{2}{3}$$

Because $g''(1)$ is positive, you know that g is concave up at $x = 1$ and, therefore, that there's a local min there. The Second Derivative Test is no help where the first derivative is undefined (where $x = 0$), so you've got to use the First Derivative Test for that critical number.

Finding Absolute Extrema on a Closed Interval

Every function that's continuous on a closed interval has an absolute maximum and an absolute minimum value in that interval — a highest and lowest point — though, as you see in the following example, there can be a tie for highest or lowest value.

A *closed* interval like [2, 5] includes the endpoints 2 and 5. An *open* interval like (2, 5) excludes the endpoints.

Finding the absolute max and min is a snap. All you do is compute the critical numbers of the function in the given interval, determine the height of the function at each critical number, and then figure the height of the function at the two endpoints of the interval. The greatest of this set of heights is the absolute max; and the least is the absolute min. Example: Find the absolute max and min of $h(x) = \cos(2x) - 2\sin x$ in the closed interval $\left[\frac{\pi}{2}, 2\pi\right]$.

1. **Find the critical numbers of h in the *open* interval** $\left(\frac{\pi}{2}, 2\pi\right)$.

$$h(x) = \cos(2x) - 2\sin x$$
$$h'(x) = -\sin(2x) \cdot 2 - 2\cos x \qquad \text{(by the chain rule)}$$
$$0 = -2\sin(2x) - 2\cos x \qquad \text{(now divide both sides by } -2\text{)}$$
$$0 = \sin(2x) + \cos x \qquad \text{(now use a trig identity)}$$
$$0 = 2\sin x \cos x + \cos x \qquad \text{(factor out } \cos x\text{)}$$
$$0 = \cos x(2\sin x + 1)$$

$$\cos x = 0 \qquad \text{or} \qquad 2\sin x + 1 = 0$$
$$x = \frac{3\pi}{2} \qquad\qquad\qquad \sin x = -\frac{1}{2}$$
$$x = \frac{7\pi}{6}, \frac{11\pi}{6}$$

Thus, the zeros of h' are $\frac{7\pi}{6}$, $\frac{3\pi}{2}$, and $\frac{11\pi}{6}$, and because h' is defined for all input numbers, this is the complete list of critical numbers.

2. **Compute the function values (the heights) at each critical number.**

$$h\left(\frac{7\pi}{6}\right) = \cos\left(2 \cdot \frac{7\pi}{6}\right) - 2\sin\left(\frac{7\pi}{6}\right) = 0.5 - 2 \cdot (-0.5) = 1.5$$

$$h\left(\frac{3\pi}{2}\right) = \cos\left(2 \cdot \frac{3\pi}{2}\right) - 2\sin\left(\frac{3\pi}{2}\right) = -1 - 2 \cdot (-1) = 1$$

$$h\left(\frac{11\pi}{6}\right) = \cos\left(2 \cdot \frac{11\pi}{6}\right) - 2\sin\left(\frac{11\pi}{6}\right) = 0.5 - 2 \cdot (-0.5) = 1.5$$

3. **Determine the function values at the endpoints of the interval.**

$$h\left(\frac{\pi}{2}\right) = \cos\left(2 \cdot \frac{\pi}{2}\right) - 2\sin\left(\frac{\pi}{2}\right) = -1 - 2 \cdot 1 = -3$$
$$h(2\pi) = \cos(2 \cdot 2\pi) - 2\sin(2\pi) = 1 - 2 \cdot 0 = 1$$

So, from Steps 2 and 3, you've found five heights: 1.5, 1, 1.5, –3, and 1. The largest number in this list, 1.5, is the absolute max; the smallest, –3, is the absolute min.

The absolute max occurs at two points: $\left(\frac{7\pi}{6}, 1.5\right)$ and $\left(\frac{11\pi}{6}, 1.5\right)$. The absolute min occurs at one of the endpoints, $\left(\frac{\pi}{2}, -3\right)$. Figure 6-7 shows the graph of h.

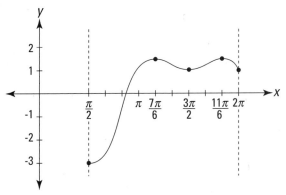

Figure 6-7: The graph of $h(x) = \cos(2x) - 2\sin x$.

Finding Absolute Extrema over a Function's Entire Domain

A function's *absolute max* and *absolute min* over its *entire domain* are the highest and lowest values of the function anywhere it's defined. A function can have an absolute max or min or both or neither. For example, the parabola $y = x^2$ has an absolute min at the point (0, 0) — the bottom of its cup shape — but no absolute max because it goes up forever to the left and the right. You could say that its absolute max is infinity if it weren't for the fact that infinity is not a number and thus it doesn't qualify as a maximum (ditto, of course, for negative infinity as a minimum).

The basic idea is this: Either a function will max out somewhere or it will go up forever to infinity. And the same idea applies to a min and going down to negative infinity.

To locate a function's absolute max and min over its domain, find the height of the function at each of its critical numbers — just like in the previous section, except that here you consider *all* the critical numbers, not just those in a given interval. The highest of these values is the absolute max unless the function rises to positive infinity somewhere, in which case you say that it has no absolute max. The lowest of these values is the absolute min, unless the function goes down to negative infinity, in which case it has no absolute min.

If a function goes up or down infinitely, it does so at its extreme right or left or at a vertical asymptote. So, your last step (after evaluating all the critical points) is to evaluate $\lim\limits_{x \to \infty} f(x)$ and $\lim\limits_{x \to -\infty} f(x)$ — the so-called *end behavior* of the function — and the limit of the function as x approaches each vertical asymptote from the left and from the right. If any of these limits equals positive infinity, then the function has no absolute max; if none equals positive infinity, then the absolute max is the function value at the highest of the critical points. And if any of these limits is negative infinity, then the function has no absolute min; if none of them equals negative infinity, then the absolute min is the function value at the lowest of the critical points.

Figure 6-8 shows a couple functions where the above method won't work. The function $f(x)$ has no absolute max despite the fact that it doesn't go up to infinity. Its max isn't 4 because it never gets to 4, and its max can't be anything less than 4, like 3.999, because it gets higher than that, say 3.9999. The function $g(x)$ has no absolute min despite the fact that it doesn't go down to negative infinity. Going left, $g(x)$ crawls along the horizontal asymptote at $y = 0$. g gets lower and lower, but it never gets as low as zero, so neither zero nor any other number can be the absolute min.

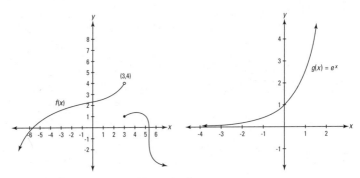

Figure 6-8: Two functions with no absolute extrema.

Concavity and Inflection Points

Look back at the function $f(x) = 3x^5 - 20x^3$ in Figure 6-2. You used the three critical numbers of f, –2, 0, and 2, to find the function's local extrema: $(-2, 64)$ and $(2, -64)$. This section investigates what happens elsewhere on this function — specifically, where the function is concave up or down and where the concavity switches (the inflection points).

The process for finding concavity and inflection points is analogous to using the First Derivative Test and the sign graph to find local extrema, except that now you use the second derivative.

1. **Find the second derivative of f.**

   $$f(x) = 3x^5 - 20x^3$$
   $$f'(x) = 15x^4 - 60x^2 \quad \text{(power rule)}$$
   $$f''(x) = 60x^3 - 120x \quad \text{(power rule)}$$

2. **Set the second derivative equal to zero and solve.**

   $$60x^3 - 120x = 0$$
   $$60x(x^2 - 2) = 0$$
   $$60x = 0 \quad \text{or} \quad x^2 - 2 = 0$$
   $$x = 0 \qquad\qquad x^2 = 2$$
   $$x = \pm\sqrt{2}$$

3. Determine whether the second derivative is undefined for any x-values.

$f''(x) = 60x^3 - 120x$ is defined for all real numbers, so there are no other x-values to add to the list from Step 2.

Steps 2 and 3 give you what you could call second derivative critical numbers of f because they are analogous to the critical numbers of f that you find using the first derivative. But, as far as I'm aware, this set of numbers has no special name. In any event, the important thing to know is that this list is made up of the zeros of f'' plus any x-values where f'' is undefined.

4. Plot these numbers on a number line and test the regions with the second derivative.

Use -2, -1, 1, and 2 as test numbers (Figure 6-9 shows the sign graph).

$$f''(-2) = -240$$
$$f''(-1) = 60$$
$$f''(1) = -60$$
$$f''(2) = 240$$

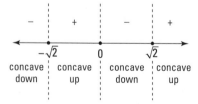

Figure 6-9: A second derivative sign graph for $f(x) = 3x^5 - 20x^3$.

A positive sign on this sign graph tells you that the function is concave up in that interval; negative means concave down. The function has an inflection point (usually) at any x-value where the signs switch from positive to negative or vice versa.

Because the signs switch at $-\sqrt{2}$, 0, and $\sqrt{2}$ and because these three numbers are zeros of f'', inflection points occur at these x-values. But in a problem where the signs switch at a number where f'' is undefined, you have to check one more thing before you can conclude that there's an inflection point there. An inflection point exists at a given x-value only if there's a tangent line to the function at that number. This is the case wherever the first derivative exists or where there's a vertical tangent.

5. **Plug these three x-values into f to obtain the function values of the three inflection points.**

$$f\left(-\sqrt{2}\right) \approx 39.6$$
$$f(0) = 0$$
$$f\left(\sqrt{2}\right) \approx -39.6$$

The square root of two equals about 1.4, so there are inflection points at about $(-1.4,\ 39.6)$, $(0, 0)$, and about $(1.4, -39.6)$.

Graphs of Derivatives

You can learn a lot about functions and their derivatives by comparing the important features of their graphs. Let's keep going with the same function $f(x) = 3x^5 - 20x^3$; travel along f from left to right (see Figure 6-10), pausing to observe its points of interest and what's happening to the graph of $f'(x) = 15x^4 - 60x^2$ at the same points. But first a (long) warning:

Looking at the graph of f' in Figure 6-10, or the graph of any derivative, you may need to remind yourself frequently that "This is the *derivative* I'm looking at, *not* the function!" You've looked at so many graphs of functions over the years that when you start looking at graphs of derivatives, you can easily lapse into thinking of them as regular functions. You might, for instance, look at an interval that's going up on the graph of a derivative and mistakenly conclude that the original function must also be going up in the same interval — an easy mistake to make. You know that the first derivative is the same thing as slope. So, when you see the graph of the first derivative going up, you may think, "Oh, the first derivative (the slope) is going up, and when the slope goes up that's like going up a hill, so the original function must be rising." This sounds reasonable because, *loosely* speaking, you can describe the front side of a hill as a slope that's going up, increasing. But *mathematically*

speaking, the front side of a hill has a *positive* slope, not necessarily an *increasing* slope. So, where a function is increasing, the graph of its derivative will be *positive,* but the derivative might be going up or down. Say you're going up a hill. As you approach the top, you're still going *up,* but the *slope* (the steepness) is going *down.* It might be 3, then 2, then 1, and then at the top the slope is zero. In such an interval, the graph of the function is *increasing,* but the graph of its derivative is *decreasing.* Got that?

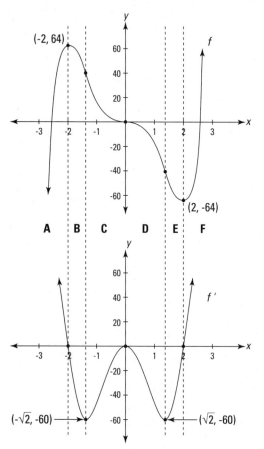

Figure 6-10: $f(x) = 3x^5 - 20x^3$ and its first derivative, $f'(x) = 15x^4 - 60x^2$.

Okay, let's get back to *f* and its derivative, shown in Figure 6-10. Remember, before the warning we were traveling along f from left to right. Beginning on the left, *f* increases until the local max at $(-2, 64)$. It's going up, so its slope is *positive,* but *f* is getting less steep so its slope is *decreasing* — the slope decreases until it becomes zero at the peak. This corresponds to the graph of *f'* (the slope) which is *positive* (because it's above the *x*-axis) but *decreasing* as it goes down to the point $(-2, 0)$.

Ready for some rules about how the graph of a function compares to the graph of its derivative? Well, ready or not, here you go:

✔ An *increasing* interval on a function corresponds to an interval on the graph of its derivative that's *positive* (or *zero* for one point if the function has a horizontal inflection point). In other words, a function's increasing interval corresponds to a part of the derivative graph that's above the *x*-axis (or that touches the axis for a single point in the case of a horizontal inflection point). See intervals A and F in Figure 6-10.

✔ A local *max* on the graph of a function corresponds to a *zero* (or *x*-intercept) on an interval of the graph of its derivative that crosses the *x*-axis going *down*.

When you look at points on a derivative graph, don't forget that the *y*-coordinate of a point — like $(-2, 0)$ — on a graph of a first derivative tells you the *slope* of the original function, not its height. Think of the *y*-axis on a first derivative graph as the *slope*-axis or the *m*-axis.

✔ A *decreasing* interval on a function corresponds to a *negative* interval on the graph of the derivative (or *zero* for one point if the function has a horizontal inflection point). The negative interval on the derivative graph is below the *x*-axis (or in the case of a horizontal inflection point, the derivative graph touches the *x*-axis at a single point). See intervals B, C, D, and E in Figure 6-10, where *f* goes down all the way to the local min at $(2, -64)$ and where *f'* is negative — except for the point $(0, 0)$ — until it gets to $(2, 0)$.

✔ A local *min* on the graph of a function corresponds to a *zero* (or *x*-intercept) on an interval of the graph of its derivative that crosses the *x*-axis going *up*.

Now retrace your steps and look at the concavity and inflection points of *f* in Figure 6-10. First, consider intervals A and B in the figure. Starting from the left again, the graph of *f* is concave down — a *decreasing* slope — until it gets to the inflection point at about $(-1.4, 39.6)$. So, the graph of *f'* decreases until it bottoms out at about $(-1.4, -60)$. These coordinates tell you that the inflection point at −1.4 on *f* has a slope of −60. Note that the inflection point at $(-1.4, 39.6)$ is the steepest point on that stretch of the function, but it has the *smallest* slope because its slope is a larger *negative* than the slope at any nearby point.

Between $(-1.4, 39.6)$ and the next inflection point at $(0, 0)$, *f* is concave up, which means the same thing as an *increasing* slope. So the graph of *f'* increases from about −1.4 to where it hits a local max at $(0, 0)$. See interval C in Figure 6-10.

Time for a couple more rules:

- ✔ A concave *down* interval on the graph of a function corresponds to a *decreasing* interval on the graph of its derivative — intervals A, B, and D in Figure 6-10. And a concave *up* interval on the function corresponds to an *increasing* interval on the derivative — intervals C, E, and F.

- ✔ An *inflection point* on a function (except for a vertical inflection point where the derivative is undefined) corresponds to a *local extremum* on the graph of its derivative. An inflection point of *minimum* slope corresponds to a local *min* on the derivative graph; an inflection point of *maximum* slope corresponds to a local *max* on the derivative graph.

After $(0, 0)$, *f* is concave down till the inflection point at about $(1.4, -39.6)$ — this corresponds to the decreasing section of *f'* from $(0, 0)$ to its min at $(1.4, -60)$ — interval D in Figure 6-10. Finally, *f* is concave up the rest of the way, which corresponds to the increasing section of *f'* beginning at $(1.4, -60)$ — intervals E and F in Figure 6-10.

The Mean Value Theorem

You don't need the Mean Value Theorem for much, but it's a famous and important theorem, so you really should learn it. Look at Figure 6-11 and see if you can make sense of the mumbo jumbo beneath it.

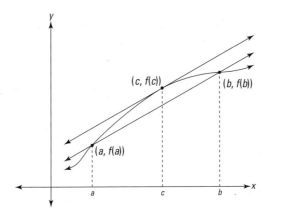

Figure 6-11: An illustration of the Mean Value Theorem.

The Mean Value Theorem: If f is continuous on the closed interval $[a, b]$ and differentiable on the open interval (a, b), then there exists at least one number c in (a, b) such that

$$f'(c) = \frac{f(b) - f(a)}{b - a}$$

Okay, so here's what the theorem means. The secant line connecting points $(a, f(a))$ and $(b, f(b))$ in Figure 6-11 has a slope given by the slope formula:

$$Slope = \frac{y_2 - y_1}{x_2 - x_1}$$
$$= \frac{f(b) - f(a)}{b - a}$$

Note that this is the same as the right side of the equation in the mean value theorem. The derivative at a point is the same thing as the slope of the tangent line at that point, so the theorem just says that there must be at least one point between a and b where the slope of the tangent is the same as the slope of the secant line from a to b. The result is parallel lines like you see in Figure 6-11.

Chapter 7

Differentiation Problems

● ●

● ●

*I*n this chapter I finally get around to showing you how to use calculus to solve some practical problems.

Optimization Problems

One practical use of differentiation is finding the maximum or minimum value of a real-world function: the maximum output of a factory or range of a missile, the minimum time to accomplish some task, and so on. Here's an example.

The maximum area of a corral

A rancher can afford 300 feet of fencing to build a corral that's divided into two equal rectangles. See Figure 7-1. What dimensions will maximize the corral's area?

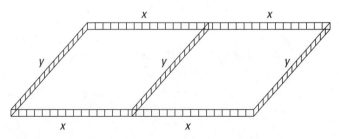

Figure 7-1: Calculus for cowboys — maximizing a corral.

1.a. Express the thing you want maximized, the area, as a function of the two unknowns, x and y.

$$A = l \cdot w$$
$$= (2x)(y)$$

In some optimization problems, you can write the thing you're trying to maximize or minimize as a function of one variable — which is always what you want. But here, the area is a function of two variables, so Step 1 has two additional substeps.

1.b. Use the given information to relate the two unknowns to each other.

The fencing is used for seven sections, thus

$$300 = x + x + x + x + y + y + y$$
$$= 4x + 3y$$

1.c. Solve this equation for y and plug the result into the y in the equation from Step 1.a. This gives you what you need — a function of one variable.

$$4x + 3y = 300$$
$$y = \frac{300 - 4x}{3}$$
$$y = 100 - \frac{4}{3}x$$
$$A = (2x)(y)$$
$$A(x) = (2x)\left(100 - \frac{4}{3}x\right)$$
$$A(x) = 200x - \frac{8}{3}x^2$$

2. Determine the domain of the function.

You can't have a negative length of fence, so x can't be negative, and the most x can be is 300 divided by 4, or 75. Thus, $0 \le x \le 75$. You now want to find the maximum value of $A(x)$ in this interval. You use the method from Chapter 6's "Finding Absolute Extrema on a Closed Interval."

3. Find the critical numbers of $A(x)$ in the open interval (0, 75) by setting its derivative equal to zero

and solving. And don't forget to check for numbers where the derivative is undefined.

$$A(x) = 200x - \frac{8}{3}x^2$$

$$A'(x) = 200 - \frac{16}{3}x \quad \text{(by the power rule)}$$

$$200 - \frac{16}{3}x = 0$$

$$-\frac{16}{3}x = -200$$

$$x = -200 \cdot \left(-\frac{3}{16}\right)$$

$$= 37.5$$

Because A' is defined for all x-values, 37.5 is the only critical number.

4. **Evaluate the function at the critical number, 37.5, and at the endpoints of the interval, 0 and 75.**

$$A(0) = 0$$
$$A(37.5) = 3,750$$
$$A(75) = 0$$

The min or max you want isn't often at an endpoint, but it can be, so be sure to evaluate the function at the interval's two endpoints.

The maximum value in the interval is 3,750, and thus, an x-value of 37.5 feet maximizes the corral's area. The length is $2x$, or 75 feet. The width is y, which equals $100 - \frac{4}{3}x$. Plugging in 37.5 gives you $100 - \frac{4}{3}(37.5)$, or 50 feet. So the rancher will build a 75'×50' corral with an area of 3,750 square feet.

Position, Velocity, and Acceleration

Every time you get in your car, you witness differentiation first hand. Your speed is the first derivative of your position. And when you accelerate or decelerate, you experience a second derivative.

If a function gives the position of something as a function of time, the first derivative gives its velocity, and the second derivative gives its acceleration. So, you differentiate *position* to get *velocity,* and you differentiate *velocity* to get *acceleration.*

Here's an example: A yo-yo moves straight up and down. Its height above the ground, as a function of time, is given by the function $H(t) = t^3 - 6t^2 + 5t + 30$, where t is in seconds and $H(t)$ is in inches. At $t = 0$, it's 30 inches above the ground, and after 4 seconds, it's at a height of 18 inches. Velocity, $V(t)$, is the derivative of position (height in this problem), and acceleration, $A(t)$, is the derivative of velocity. Thus —

$$H(t) = t^3 - 6t^2 + 5t + 30$$
$$V(t) = H'(t) = 3t^2 - 12t + 5$$
$$A(t) = V'(t) = H''(t) = 6t - 12$$

Take a look at the graphs of these three functions in Figure 7-2.

Using the three functions and their graphs, I want to discuss several things about the yo-yo's motion:

- ✔ Maximum and minimum height
- ✔ Maximum, minimum, and average velocity
- ✔ Total displacement
- ✔ Maximum, minimum, and average speed
- ✔ Total distance traveled
- ✔ Positive and negative acceleration
- ✔ Speeding up and slowing down

That's a lot to cover, so I'm going to cut some corners — like not always checking endpoints when looking for extrema if it's obvious they don't occur there. But before discussing the bullet points, let's go over a few things about velocity, speed, and acceleration.

Velocity versus speed

Your friends won't complain — or even notice — if you use the words "velocity" and "speed" interchangeably, but your friendly mathematician *will* complain. Here's the difference.

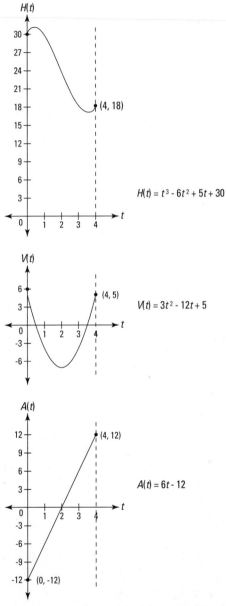

$H(t) = t^3 - 6t^2 + 5t + 30$

$V(t) = 3t^2 - 12t + 5$

$A(t) = 6t - 12$

Figure 7-2: The graphs of the yo-yo's height, velocity, and acceleration functions from 0 to 4 seconds.

For the velocity function in Figure 7-2, *upward* motion is defined as a *positive* velocity, and *downward* motion is a *negative* velocity — this is the standard way velocity is treated in most calculus and physics problems. (If the motion is horizontal, going right is a positive velocity and going left is a negative velocity.)

Speed, on the other hand, is always positive (or zero). If a car goes by at 50 *mph,* for instance, you say its speed is 50, and you mean *positive* 50, regardless of whether it's going to the right or the left. For velocity, the direction matters; for speed it doesn't. In everyday life, speed is a simpler idea than velocity because it agrees with common sense. But in calculus, speed is actually the trickier idea because it doesn't fit nicely into the three-function scheme shown in Figure 7-2.

 You've got to keep the velocity-speed distinction in mind when analyzing velocity and acceleration. For example, if an object is going down (or to the left) faster and faster, its speed is increasing, but its velocity is *decreasing* because its velocity is becoming a bigger negative (and bigger negatives are smaller numbers). This seems weird, but that's the way it works. And here's another strange thing: Acceleration is defined as the rate of change of velocity, not speed. So, if an object is slowing down while going in the downward direction, and thus has an *increasing* velocity — because the velocity is becoming a smaller negative — the object has a *positive* acceleration. In everyday English, you'd say the object is decelerating (slowing down), but in calculus class, you say that the object has a negative velocity and a positive acceleration. (By the way, "deceleration" isn't exactly a technical term, so you should probably avoid it in calculus class. It's best to use the following vocabulary: "positive acceleration," "negative acceleration," "speeding up," and "slowing down." More on these things later.)

Maximum and minimum height

The maximum and minimum of $H(t)$ occur at the local extrema seen in Figure 7-2. To locate them, set the derivative of $H(t)$, that's $V(t)$, equal to zero and solve.

$$H'(t) = V(t) = 3t^2 - 12t + 5$$

$$0 = 3t^2 - 12t + 5$$

$$t = \frac{-(-12) \pm \sqrt{(-12)^2 - 4(3)(5)}}{2 \cdot 3} \qquad \text{(quadratic formula)}$$

$$= \frac{12 \pm \sqrt{84}}{6}$$

$$= \frac{12 \pm 2\sqrt{21}}{6}$$

$$= \frac{6 \pm \sqrt{21}}{3} \approx 0.47 \text{ or } 3.53$$

These two numbers are the zeros of $V(t)$ and the t-coordinates (*time*-coordinates) of the max and min of $H(t)$, which you can see in Figure 7-2. In other words, these are the *times* when the yo-yo reaches its maximum and minimum heights. Plug these numbers into $H(t)$ to obtain the heights:

$$H(0.47) \approx 31.1$$

$$H(3.53) \approx 16.9$$

So the yo-yo gets as high as about 31.1 inches above the ground at $t \approx 0.47$ second and as low as about 16.9 inches at $t \approx 3.53$ seconds.

Velocity and displacement

Again, speed is always positive, but going down (or left) is a negative velocity. The connection between *displacement* and *distance traveled* is similar: Distance traveled is always positive (or zero), but going down (or left) is a *negative* displacement. The basic idea is this: If you drive to a store 1 mile away — taking the scenic route and clocking 3 miles on your odometer — your total distance traveled is 3 miles, but your displacement is just 1 mile.

Total displacement

Total displacement equals final position minus initial position. So, because the yo-yo starts at a height of 30 and ends at a height of 18,

$$\text{Total displacement} = 18 - 30 = -12$$

This is negative because the net movement is *downward*.

Average velocity

Average velocity is given by total displacement divided by elapsed time. Thus,

$$Average\ velocity = -\frac{12}{4} = -3$$

This negative answer tells you that the yo-yo, on average, is going *down* 3 *inches per second.*

Maximum and minimum velocity

To determine the yo-yo's maximum and minimum velocity during the interval from 0 to 4 seconds, set the derivative of $V(t)$, that's $A(t)$, equal to zero and solve:

$$V'(t) = A(t) = 6t - 12$$
$$6t - 12 = 0$$
$$6t = 12$$
$$t = 2$$

Now, evaluate $V(t)$ at the critical number, 2, and at the interval's endpoints, 0 and 4:

$$V(0) = 5$$
$$V(2) = -7$$
$$V(4) = 5$$

The yo-yo has a maximum velocity of 5 *inches per second* twice — at the beginning and the end of the interval. It reaches a minimum velocity of –7 *inches per second* at $t = 2$ seconds.

Speed and distance traveled

Unlike *velocity* and *displacement,* which have technical definitions, *speed* and *distance traveled* have commonsense meanings. *Speed,* of course, is what you read on your speedometer, and you can read *distance traveled* on your odometer or "tripometer."

Total distance traveled

To determine total distance, add the distances traveled on each leg of the yo-yo's trip: the up leg, the down leg, and the second up leg.

First, the yo-yo goes up from a height of 30 inches to about 31.1 inches (where the first turn-around point is), a distance of about 1.1 inches. Next, it goes down from about 31.1 to about 16.9 (the height of the second turn-around point). That's 31.1 – 16.9, or about 14.2 inches. Finally, it goes up again from about 16.9 to its final height of 18 inches, another 1.1 inches. Add the three distances to get total distance travelled: ~ 1.1+ ~ 14.2+ ~ 1.1 ≈ 16.4 inches.

Average speed

The yo-yo's average speed is given by the total distance traveled divided by the elapsed time. Thus,

$$Average\ speed \approx \frac{16.4}{4} \approx 4.1\ inches\ per\ second$$

Maximum and minimum speed

You previously determined the yo-yo's maximum velocity (5 *inches per second*) and its minimum velocity (–7 *inches per second*). A velocity of –7 is a speed of 7, so that's the yo-yo's maximum speed. Its minimum speed of zero occurs at the two turnaround points.

For a continuous *velocity* function, the *minimum speed* is zero whenever the maximum and minimum velocities are of opposite signs or when one of them is zero. When the maximum and minimum velocities are both positive or both negative, then the *minimum* speed is the *lesser* of the absolute values of the maximum and minimum velocities. In all cases, the *maximum* speed is the *greater* of the absolute values of the maximum and minimum velocities. Is that a mouthful or what?

Acceleration

Let's go over acceleration: Put your pedal to the metal.

Positive and negative acceleration

The graph of the acceleration function at the bottom of Figure 7-2 is a simple line, $A(t) = 6t - 12$. It's easy to see that the acceleration of the yo-yo goes from a minimum of $-12 \frac{inches\ per\ second}{second}$ at $t = 0$ seconds to a maximum of

$12 \dfrac{inches\ per\ second}{second}$ at $t = 4$ seconds, and that the acceleration
is zero at $t = 2$ when the yo-yo reaches its minimum velocity (and
maximum speed). When the acceleration is *negative* — on the
interval $[0, 2)$ — that means that the velocity is *decreasing*. When
the acceleration is *positive* — on the interval $(2, 4]$ — the velocity is *increasing*.

Speeding up and slowing down

Figuring out when the yo-yo is speeding up and slowing down
is probably more interesting and descriptive of its motion
than the info in the preceding section. An object is speeding
up (what we call "acceleration" in everyday speech) whenever
the velocity and the calculus acceleration are both positive
or both negative. And an object is slowing down (what we call
"deceleration") when the velocity and the calculus acceleration are of opposite signs.

Look at all three graphs in Figure 7-2 again. From $t = 0$ to about
$t = 0.47$ (when velocity is zero), the velocity is positive and the
acceleration is negative, so the yo-yo is slowing down (till it
reaches its maximum height). When $t = 0$, the deceleration is
greatest $(12 \dfrac{inches\ per\ second}{second}$; the graph shows *negative* 12,
but I'm calling it positive 12 because I'm calling it a deceleration, get it?) From about $t = 0.47$ to $t = 2$, both velocity and
acceleration are negative, so the yo-yo is speeding up. From
$t = 2$ to about $t = 3.53$, velocity is positive and acceleration is
negative, so the yo-yo is slowing down again (till it bottoms
out at its lowest height). Finally, from about $t = 3.53$ to $t = 4$,
both velocity and acceleration are positive, so the yo-yo is
speeding up again. The yo-yo reaches its greatest acceleration
of $12 \dfrac{inches\ per\ second}{second}$ at $t = 4$ seconds.

Tying it all together

Note the following connections among the three graphs in
Figure 7-2. The *negative* section on the graph of $A(t)$ — from
$t = 0$ to $t = 2$ — corresponds to a *decreasing* section of the graph
of $V(t)$ and a *concave down* section of the graph of $H(t)$. The
positive interval on the graph of $A(t)$ — from $t = 2$ to $t = 4$ —
corresponds to an *increasing* interval on the graph of $V(t)$ and
a *concave up* interval on the graph of $H(t)$. When $t = 2$ seconds,

$A(t)$ has a zero, $V(t)$ has a *local minimum*, and $H(t)$ has an *inflection point.*

Related Rates

Say you're filling up your swimming pool; you know how fast water is coming out of your hose, and you want to calculate how fast the water level in the pool is rising. You know one rate (how fast the water is pouring in), and you want to determine another rate (how fast the water level is rising). These rates are called *related rates* because one depends on the other — the faster the water is poured in, the faster the water level rises. In a typical related rates problem, the rate or rates you're given are unchanging, but the rate you have to figure out is changing with time. You have to determine this rate at one particular point in time.

A calculus crossroads

One car leaves an intersection traveling north at 50 *mph*; another is driving west toward the intersection at 40 *mph*. At one point, the north-bound car is three-tenths of a mile north of the intersection and the west-bound car is four-tenths of a mile east of it. At this point, how fast is the distance between the cars changing?

1. **Draw a diagram and label it with any *unchanging* measurements (there are none in this particular problem), and make sure to assign a variable to anything in the problem that's *changing* (unless it's irrelevant to the problem). See Figure 7-3.**

   Note that variables have been assigned to the distance between Car A and the intersection, the distance between Car B and the intersection, and the distance between the two cars, because those are *changing* distances. The 0.3 and 0.4 are in parentheses to emphasize that they're *not* unchanging measurements; they're the distances at one particular point in time.

   In related rates problems, it's important to distinguish between what is changing and what is *not* changing.

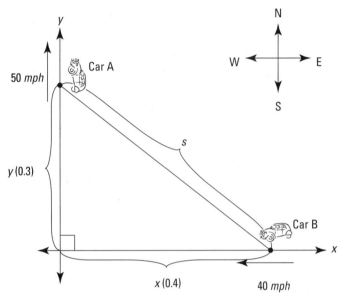

Figure 7-3: Calculus — it's a drive in the country.

You may run across a similar problem in your calculus textbook. It involves a ladder leaning against and
sliding down a wall. The diagram for this problem
would be very similar to Figure 7-3 except that the
y-axis would represent the wall, the *x*-axis would be
the ground, and the diagonal line would be the ladder.
Despite the similarities between these two problems,
there's an important difference: The distance between
the cars is *changing*, so the diagonal line in Figure 7-3
is labeled with a variable, *s*. A ladder, on the other
hand, has a *fixed* length, so the diagonal line in the diagram for the ladder problem would be labeled with a
number, not a variable.

2. **List all given rates and the rate you're asked to
   determine as derivatives with respect to time.**

   As Car A travels north, *y* is growing at 50 *miles per
   hour.* That's a rate, a change in distance per change in
   time. So,

   $$\frac{change\ in\ distance\ in\ y\ direction}{change\ in\ time} = \frac{dy}{dt} = 50\ mph$$

As Car B travels west, x is *shrinking* at 40 *miles per hour*. That's a *negative* rate:

$$\frac{\text{change in distance in } x \text{ direction}}{\text{change in time}} = \frac{dx}{dt} = -40 \text{ mph}$$

You have to figure out how fast s is changing, so

$$\frac{\text{change in distance in } s \text{ direction}}{\text{change in time}} = \frac{ds}{dt} = ?$$

3. Write the formula that relates the variables in the problem: x, y, and s.

There's a right triangle in your diagram, so you use the Pythagorean Theorem: $x^2 + y^2 = s^2$.

The Pythagorean Theorem is used a lot in related rates problems. If there's a right triangle in your problem, it's likely that $a^2 + b^2 = c^2$ is the formula you'll need.

4. Differentiate your formula with respect to time, t.

When you differentiate in a related rates problem, all variables are treated like the ys are treated in a typical implicit differentiation problem.

$$s^2 = x^2 + y^2$$

$$2s\frac{ds}{dt} = 2x\frac{dx}{dt} + 2y\frac{dy}{dt}$$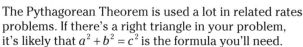
(implicit differentiation with the power rule)

5. Substitute known values for the rates and variables in the equation from Step 4, and then solve for the thing you're asked to determine, $\frac{ds}{dt}$.

$$x = 0.4, \ y = 0.3, \ \frac{dx}{dt} = -40, \ \frac{dy}{dt} = 50, \text{ and } s = \text{......}$$

There's a glitch here because you don't know s (you'll run into this glitch in some but not all related rates problems). No worries, just use the Pythagorean Theorem again.

$$s^2 = x^2 + y^2$$

$$s^2 = 0.4^2 + 0.3^2$$

$$= 0.25$$

$$s = \pm 0.5 \quad \text{(square rooting both sides)}$$

You can reject the negative answer because s obviously has a positive length. So $s = 0.5$.

Now plug everything into your equation.

$$2s\frac{ds}{dt} = 2x\frac{dx}{dt} + 2y\frac{dy}{dt}$$

$$2 \cdot 0.5 \cdot \frac{ds}{dt} = 2 \cdot 0.4 \cdot (-40) + 2 \cdot 0.3 \cdot 50$$

$$1 \cdot \frac{ds}{dt} = -32 + 30$$

$$\frac{ds}{dt} = -2$$

This negative answer means that the distance, *s*, is *decreasing*. Thus, when car A is 3 blocks north of the intersection and car B is 4 blocks east of the intersection, the distance between them is decreasing at a rate of 2 *mph*.

Be sure to differentiate (Step 4) *before* you plug the given information into the unknowns (Step 5).

Filling up a trough

Here's another garden-variety related rates problem. A trough is being filled up with swill. It's 10 feet long, and its cross-section is an isosceles triangle with a base of 2 feet and a height of 2 feet 6 inches (with the vertex at the bottom, of course). Swill's being poured in at a rate of 5 *cubic feet per minute*. When the depth of the swill is 1 foot 3 inches, how fast is the swill level rising?

1. **Draw a diagram, labeling the diagram with any** *unchanging* **measurements and assigning variables to any** *changing* **things. See Figure 7-4.**

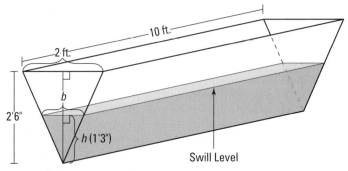

Figure 7-4: Filling a trough with swill — lunch time.

Note that Figure 7-4 shows the *unchanging* dimensions of the trough, 2 feet, 2 feet 6 inches, and 10 feet, and that these dimensions do *not* have variable names like *l* for length or *h* for height. And note that the *changing* things — the height (or depth) of the swill and the width of the surface of the swill (which gets wider as the swill gets deeper) — have variable names, *h* for height and *b* for base (I call it *base*, not *width*, because it's the base of the upside-down triangle shape made by the swill). The swill volume is also changing, so you can call that *V*, of course.

2. **List all given rates and the rate you're asked to figure out as derivatives with respect to time.**

$$\frac{dV}{dt} = 5 \text{ cubic feet per minute} \qquad \frac{dH}{dt} = ?$$

3.a. **Write down the formula that connects the variables in the problem: *V*, *h*, and *b*.**

Remember the formula for the volume of a right prism (the shape of the swill in the trough)?:

$$V = (area \ of \ base) \cdot (height)$$

Note that this "base" is the base of the prism (the whole triangle at the end of the trough), not the base of the triangle labeled *b* in Figure 7-4. Also, this "height" is the height of the prism (length of the trough), not the height labeled *h* in Figure 7-4. Sorry about the confusion. Deal with it.

The area of the triangular base equals $\frac{1}{2}bh$, and the "height" of the prism is 10 feet, so the formula becomes

$$V = \frac{1}{2}bh \cdot 10$$
$$= 5bh$$

This formula contains a variable, *b*, that you don't see in your list of derivatives in Step 2. So Step 3 has a second part — getting rid of this extra variable.

3.b. **Find an equation that relates the unwanted variable, *b*, to some other variable in the problem so you can make a substitution that leaves you with only *V* and *h*.**

The triangular face of the swill in the trough is *similar* to the triangular face of the trough itself, so the base and height of these triangles are proportional. (Recall from geometry that *similar triangles* are triangles of the same shape; their sides are proportional.) Thus,

$$\frac{b}{2} = \frac{h}{2.5}$$
$$2.5b = 2h$$
$$b = \frac{2h}{2.5}$$
$$b = 0.8h$$

Similar triangles come up a lot in related rates problems. Look for them whenever the problem involves a triangle, a triangular prism, or a cone shape.

Now substitute $0.8h$ for b in your formula from Step 3.a.

$$V = 5bh$$
$$V = 5 \cdot 0.8h \cdot h$$
$$V = 4h^2$$

4. **Differentiate this equation with respect to t.**

$$\frac{dV}{dt} = 8h\frac{dh}{dt}$$

5. **Substitute known values for the rate and variable in the equation from Step 4 and then solve.**

You know that $\frac{dV}{dt} = 5$ *cubic feet per minute*, and you want to determine $\frac{dh}{dt}$ when h equals 1 foot 3 inches, or 1.25 feet, so plug in 5 and 1.25 and solve for $\frac{dh}{dt}$:

$$5 = 8 \cdot 1.25 \cdot \frac{dh}{dt}$$
$$5 = 10 \cdot \frac{dh}{dt}$$
$$\frac{dh}{dt} = \frac{1}{2}$$

That's it. The swill's level is rising at a rate of 1/2 *foot per minute* when the swill is 1 foot 3 inches deep. Dig in.

Linear Approximation

Because ordinary functions are locally *linear* (straight) — and the further you zoom in on them, the straighter they look — a line tangent to a function is a good approximation of the function near the point of tangency. Figure 7-5 shows the graph of $f(x) = \sqrt{x}$ and a line tangent to the function at the point (9, 3). Near (9, 3), the curve and the tangent line are virtually indistinguishable.

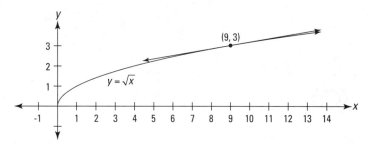

Figure 7-5: The graph of $f(x) = \sqrt{x}$ and a line tangent to the curve at (9, 3).

Determining the equation of this tangent line is a breeze. You've got a point, (9, 3), and the slope is given by the derivative of f at 9:

$$f(x) = \sqrt{x}$$
$$= x^{1/2}$$
$$f'(x) = \frac{1}{2}x^{-1/2} \quad \text{(by the power rule)}$$
$$= \frac{1}{2\sqrt{x}}$$
$$f'(9) = \frac{1}{2\sqrt{9}}$$
$$= \frac{1}{6}$$

Now just take this slope, $\frac{1}{6}$, and the point (9, 3), and plug them into the point-slope form:

$$y - y_1 = m(x - x_1)$$
$$y - 3 = \frac{1}{6}(x - 9)$$
$$y = 3 + \frac{1}{6}(x - 9)$$

That's the equation of the line tangent to $f(x) = \sqrt{x}$ at (9, 3). Are you wondering why I wrote the equation as $y = 3 + \frac{1}{6}(x - 9)$? It might seem more natural to put the 3 to the right of $\frac{1}{6}(x - 9)$, which, of course, would also be correct. And I could have simplified the equation further, writing it in $y = mx + b$ form. I explain later in this section why I wrote it the way I did.

Now, say you want to approximate the square root of 10. Because 10 is pretty close to 9, and because you can see from Figure 7-5 that $f(x)$ and its tangent line are close to each other at $x = 10$, the y-coordinate of the line at $x = 10$ is a good approximation of the function value at $x = 10$, namely $\sqrt{10}$. Just plug 10 into the line equation for your approximation:

$$y = 3 + \frac{1}{6}(x - 9)$$
$$= 3 + \frac{1}{6}(10 - 9)$$
$$= 3\frac{1}{6}$$

Thus, the square root of 10 is about $3\frac{1}{6}$. This is only about 0.004 more than the exact answer of 3.1623.

Now I can explain why I wrote the equation for the tangent line the way I did. This form makes it easier to do the computation and easier to understand what's going on when you compute an approximation. You know that the line goes through the point (9, 3), right? And you know the slope of the line is $\frac{1}{6}$. So, you can start at (9, 3) and go to the right (or left) along the line in the stair-step fashion, as shown in Figure 7-6: over 1, up $\frac{1}{6}$; over 1, up $\frac{1}{6}$; and so on.

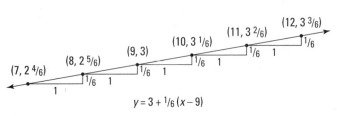

$$y = 3 + \frac{1}{6}(x - 9)$$

Figure 7-6: The linear approximation line and several of its points.

So, when you're doing an approximation, you start at a y-value of 3 and go up $\frac{1}{6}$ for each 1 you go to the right. Or if you go to the left, you go down $\frac{1}{6}$ for each 1 you go to the left. When the line equation is written in the above form, the computation of an approximation parallels this stair-step scheme.

Figure 7-6 shows the approximate values for the square roots of 7, 8, 10, 11, and 12. Here's how you come up with these values: To get to 8, for example, from (9, 3), you go 1 to the left, so you go down $\frac{1}{6}$ to $2\frac{5}{6}$; or to get to 11 from (9, 3), you go *two* to the right, so you go up *two*-sixths to $3\frac{2}{6}$ or $3\frac{1}{3}$. (If you go to the right *one half* to $9\frac{1}{2}$, you go up *half* of a sixth, that's a twelfth, to $3\frac{1}{12}$ — the approximate square root of $9\frac{1}{2}$.)

Below are the errors for the approximations shown in Figure 7-6. Note that the errors grow as you get further from the point of tangency (9, 3); also, the errors grow faster going down from (9, 3) than going up from (9, 3) — errors often grow faster in one direction than the other with linear approximations.

$$\sqrt{7}:\ \ 0.8\%\ \text{error}$$
$$\sqrt{8}:\ \ 0.2\%\ \text{error}$$
$$\sqrt{10}:\ \ 0.1\%\ \text{error}$$
$$\sqrt{11}:\ \ 0.5\%\ \text{error}$$
$$\sqrt{12}:\ \ 1.0\%\ \text{error}$$

Linear Approximation Equation: Here's the general form for the equation of the tangent line for a linear approximation. The values of a function $f(x)$ can be approximated by the values of the tangent line $l(x)$ near the point of tangency, $(x_0, f(x_0))$, where

$$l(x) = f(x_0) + f'(x_0)(x - x_0)$$

This is less complicated than it looks. It's just the gussied-up calculus version of the point-slope equation of a line you've known since Algebra I, $y - y_1 = m(x - x_1)$, with the y_1 moved to the right side:

$$y = y_1 + m(x - x_1)$$

This algebra equation and the above equation for $l(x)$ differ only in the symbols used; the *meaning* of both equations — term for term — is identical. And notice how both equations resemble the equation of the tangent line in Figure 7-6.

Whenever possible, try to see the basic algebra or geometry concepts at the heart of fancy-looking calculus concepts.

Chapter 8

Introduction to Integration

- -

In This Chapter

▶ Integrating — adding it all up

▶ Approximating areas

▶ Using the definite integral to get exact areas

- -

*I*f you're still reading, I presume you survived differentiation (Chapters 4–7). Now you begin the second major calculus topic: integration. Just as two simple ideas lie at the heart of differentiation — *rate* (like *miles per hour*) and the *slope* of a curve — integration can also be understood in terms of two simple ideas: *adding up* small pieces of something and the *area* under a curve.

Integration: Just Fancy Addition

Say you want to determine the volume of the lamp's base in Figure 8-1. Because no formula for the volume of such a weird shape exists, you can't calculate the volume directly. You can, however, calculate the volume with integration. Imagine that the base is cut up into thin, horizontal slices, as shown on the right in Figure 8-1.

Do you see how each slice is shaped like a thin pancake? Now, because there *is* a formula for the volume of a pancake (a pancake is just a very short cylinder), you can determine the total volume of the base by simply calculating the volume of each pancake-shaped slice and then adding up the volumes. That's integration in a nutshell.

Figure 8-1: A lamp with a curvy base and the base
cut into thin horizontal slices.

What makes integration one of the great achievements in
the history of mathematics is that — to stay with the lamp
example — it gives you the *exact* volume of the lamp's base by
sort of cutting it into an *infinite* number of *infinitely* thin slices.
If you cut the lamp into fewer than an infinite number of slices,
you can get only a very good approximation of the volume
because each slice will have a weird, curved edge which would
cause a small error.

Integration has an elegant symbol: $\int$. You've probably seen it
before. You can think of the integration symbol as an elongated
S for "sum up." So, for our lamp problem, you can write

$$\int_{\text{bottom}}^{\text{top}} dB = B$$

where dB means a little bit of the base — actually an infinitely
small piece. So the equation just means that if you sum up
all the little pieces of the base from the *bottom* to the *top*, the
result is B, the volume of the whole base.

This is a bit oversimplified — I can hear the siren of the math
police now — but it's a good way to think about integration.
By the way, thinking of dB as a little or infinitesimal piece of B
is an idea you saw before with differentiation (see Chapter 4),
where the derivative or slope, $\frac{dy}{dx}$, is equal to the ratio of a
little bit of y to a little bit of x, as you shrink the slope stair
step down to an infinitesimal size — see Figure 8-2.

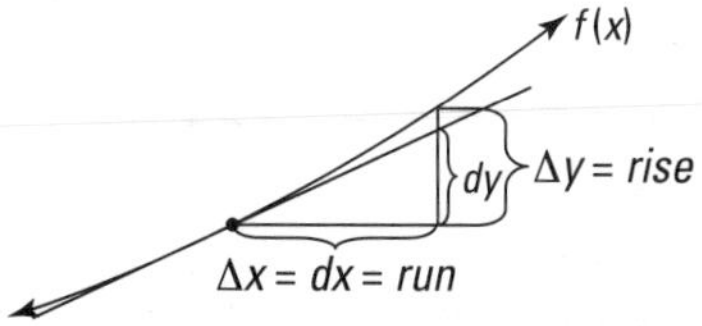

Figure 8-2: In the limit, $\dfrac{dy}{dx} = \dfrac{\text{a little bit of } y}{\text{a little bit of } x} = \dfrac{\Delta y}{\Delta x} = \dfrac{rise}{run} = \text{slope}$.

So, whenever you see something like

$$\int_a^b \text{little piece of mumbo jumbo}$$

it just means that you add up all the little pieces of the mumbo jumbo from *a* to *b* to get the total of all of the mumbo jumbo from *a* to *b*. Or you might see something like

$$\int_{t=0}^{t=20} \text{little piece of distance}$$

which means add up the little pieces of distance traveled between 0 and 20 seconds to get the total distance traveled in that time span.

To sum up — that's a pun! — the mathematical expression to the right of the integration symbol always stands for a little bit of something, and integrating such an expression means to add up all the little pieces between some starting point and some ending point.

Finding the Area under a Curve

The most fundamental meaning of integration is *to add up*. And when you depict integration on a graph, you can see the adding up process as a summing up of little bits of area to arrive at the total area under a curve. Consider Figure 8-3.

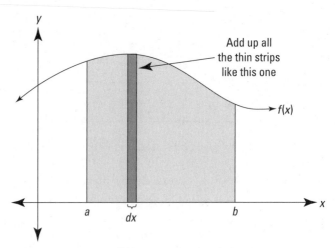

Figure 8-3: Integrating *f(x)* from *a* to *b* means finding the area under the curve between *a* and *b*.

The shaded area in Figure 8-3 can be calculated with the following integral:

$$\int_a^b f(x)\,dx$$

Look at the thin rectangle in Figure 8-3. It has a height of $f(x)$ and a width of dx (a little bit of x), so its area (*length* × *width*, of course) is given by $f(x)dx$. The above integral tells you to add up the areas of all the narrow rectangular strips between a and b under the curve $f(x)$. As the strips get narrower and narrower, you get a better and better estimate of the area. The power of integration lies in the fact that it gives you the *exact* area by sort of adding up an infinite number of infinitely thin rectangles.

If you're doing a problem where both the x and y axes are labeled in a unit of length, say, *feet,* then each thin rectangle measures so many feet by so many feet, and its area — *length* × *width* — is some number of *square feet.* In this case, when you integrate to get the total area under the curve between a and b, your final answer will be an amount of — what else? — area. But you can use this adding-up-areas-of-rectangles scheme to add up tiny bits of anything — distance, volume, or energy, for example. In other words, the area under the curve doesn't have to stand for an actual area.

If, for example, the units on the *x*-axis are *hours* and the *y*-axis is labeled in *miles per hour,* then, because *rate × time = distance,* the area of each rectangle represents an amount of distance, and the total area gives you the total distance traveled during the given time interval. Or if the *x*-axis is labeled in *hours* and the *y*-axis in *kilowatts* of electrical power — in which case the curve gives power usage as a function of time — then the area of each rectangular strip (*kilowatts × hours*) represents a number of *kilowatt-hours* of energy. In that case, the total area under the curve gives you the total number of kilowatt-hours of energy consumption between two points in time.

Dealing with negative area

In the examples involving volume, distance, and energy (see previous two sections), you're adding up *positive* bits of something. This is common with practical problems because you can't, say, have a negative volume of water or use a negative number of kilowatt-hours. However, you will sometimes integrate functions that go into the negatives — that's below the *x*-axis. Here's what you do.

When using integration to calculate area, area *below* the *x*-axis counts as *negative* area. The total area between *a* and *b* for some curve $f(x)$ — given by the integral $\int_a^b f(x)dx$ — is really a *net* area where the total area below the *x*-axis (and above the curve) is subtracted from the total area above the *x*-axis (and below the curve).

Approximating Area

Before explaining how to calculate exact areas, I want to show you how to approximate areas. The approximation method is useful not only because it lays the groundwork for the exact method — integration — but because for some curves, integration is impossible, and an approximation of area is the best you can do.

Approximating area with left sums

Say you want the exact area under the curve $f(x) = x^2 + 1$ between 0 and 3. See the shaded area on the graph on the left in Figure 8-4.

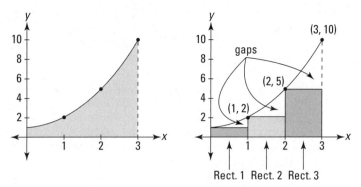

Figure 8-4: The exact area under $f(x) = x^2 + 1$ between 0 and 3 (left) is approximated by the area of three rectangles (right).

You can get a rough estimate of the area by drawing three rectangles under the curve, as shown on the right in Figure 8-4, and then adding up their areas.

The rectangles in Figure 8-4 represent a so-called *left sum* because the upper *left* corner of each rectangle touches the curve. Each rectangle has a width of 1 and the height of each is given by the height of the function at the rectangle's left edge. So, rectangle number 1 has a height of $f(0) = 0^2 + 1 = 1$; its area (*height* $\times$ *width*) is thus 1×1, or 1. Rectangle 2 has a height of $f(1) = 1^2 + 1 = 2$, so its area is 2×1, or 2. And rectangle 3 has a height of $f(2) = 2^2 + 1 = 5$, so its area is 5×1, or 5. Adding the areas gives you $1 + 2 + 5$, or 8. This is an underestimate of the total area under the curve because of the gaps between the rectangles and the curve.

For a better estimate, double the number of rectangles to six. Figure 8-5 shows six "left" rectangles under the curve and also how they begin to fill up the three gaps you see in Figure 8-4.

See the three small, darker shaded rectangles in the graph on the right in Figure 8-5? They sit on top of the three rectangles from Figure 8-4, and they represent how much the area estimate has improved by using six rectangles instead of three.

Now total up the areas of the six rectangles. Each has a width of 0.5 and the heights are $f(0)$, $f(0.5)$, $f(1)$, $f(1.5)$, and so on. I'll spare you the math. Here's the total: $0.5 + 0.625 + 1 + 1.625 + 2.5 + 3.625 = 9.875$. This is a better estimate, but it's still an underestimate because of the six small gaps you can see on the left graph in Figure 8-5.

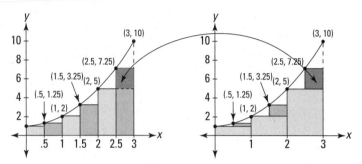

Figure 8-5: Six "left" rectangles approximate the area under $f(x) = x^2 + 1$.

Table 8-1 shows the area estimates given by 3, 6, 12, 24, 48, 96, 192, and 384 rectangles. You don't have to double the rectangles each time — you can use any number of rectangles you want. I just like the doubling scheme because, with each doubling, the gaps are plugged up more and more in the way shown in Figure 8-5.

Table 8-1 Estimates of the Area under $f(x) = x^2 + 1$ Given by Increasing Numbers of "Left" Rectangles

Number of Rectangles	Area Estimate
3	8
6	9.875
12	~10.906
24	~11.445
48	~11.721
96	~11.860
192	~11.930
384	~11.965

The Left Rectangle Rule: You can approximate the exact area under a curve between a and b, $\int_a^b f(x)dx$, with a sum of *left* rectangles given by the following formula. In general, the more rectangles, the better the estimate.

$$L_n = \frac{b-a}{n}\left[f(x_0)+f(x_1)+f(x_2)+\text{............}+f(x_{n-1})\right]$$

Where n is the number of rectangles, $\frac{b-a}{n}$ is the width of each one, and the function values are the heights of the rectangles.

Look back to the six rectangles shown in Figure 8-5. The width of each rectangle equals the length of the total span from 0 to 3 (which of course is $3-0$, or 3) divided by the number of rectangles, 6. That's what the $\frac{b-a}{n}$ does in the formula.

What about those xs with the subscripts? The x-coordinate of the *left* edge of rectangle 1 in Figure 8-5 is x_0, the *right* edge of rectangle 1 (the same as the left edge of rectangle 2) is at x_1, the right edge of rectangle 2 is at x_2, the right edge of rectangle 3 is at x_3, and so on up to the right edge of rectangle 6, at x_6. For the six rectangles in Figure 8-5, x_0 is 0, x_1 is 0.5, x_2 is 1, x_3 is 1.5, x_4 is 2, x_5 is 2.5, and x_6 is 3. The heights of the six left rectangles in Figure 8-5 occur at their left edges, which are at x_0 through x_5. You don't use the right edge of the last rectangle, x_6, in a left sum. That's why the list of function values in the formula stops at x_{n-1}.

Here's how to use the formula for the six rectangles in Figure 8-5:

$$L_6 = \frac{3-0}{6}\left[f(x_0)+f(x_1)+f(x_2)+f(x_3)+f(x_4)+f(x_5)\right]$$
$$= \frac{1}{2}\left[f(0)+f(0.5)+f(1)+f(1.5)+f(2)+f(2.5)\right]$$
$$= \frac{1}{2}(1+1.25+2+3.25+5+7.25)$$
$$= \frac{1}{2}(19.75)$$
$$= 9.875$$

Had I distributed the width of 1/2 to each of the heights after the third line in that solution, you'd see the sum of the areas of the rectangles — seen just before Figure 8-5. The formula uses the shortcut of first adding up the heights and then multiplying by the width.

Approximating area with right sums

Now let's estimate the same area under $f(x) = x^2 + 1$ from 0 to 3 with *right* rectangles. This method works just like the left sum method except that each rectangle is drawn so that its *right* upper corner touches the curve. See Figure 8-6.

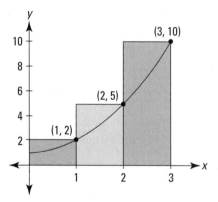

Figure 8-6: Three *right* rectangles used to approximate the area under $f(x) = x^2 + 1$.

The heights of the three rectangles in Figure 8-6 are given by the function values at their *right* edges: $f(1) = 2$, $f(2) = 5$, and $f(3) = 10$. Each rectangle has a width of 1, so the areas are 2, 5, and 10, which total 17. Table 8-2 shows the improving estimates you get with more and more right rectangles.

Table 8-2	Estimates of the Area under $f(x)=x^2+1$ Given by Increasing Numbers of "Right" Rectangles

Number of Rectangles	Area Estimate
3	17
6	14.375
12	~13.156
24	~12.570
48	~12.283
96	~12.141
192	~12.070
384	~12.035

The Right Rectangle Rule: You can approximate the exact area under a curve between a and b, $\int_a^b f(x)dx$, with a sum of *right* rectangles given by the following formula. In general, the more rectangles, the better the estimate.

$$R_n = \frac{b-a}{n}\left[f(x_1)+f(x_2)+f(x_3)+\ldots\ldots+f(x_n)\right]$$

where n is the number of rectangles, $\frac{b-a}{n}$ is the width of each, and the function values are the heights of the rectangles.

If you compare this formula to the one for a left rectangle sum, you get the complete picture about those subscripts. The two formulas are the same except for one thing. Look at the sums of the function values in both formulas. The right sum formula has one value, $f(x_n)$, that the left sum formula doesn't have, and the left sum formula has one value, $f(x_0)$, that the right sum formula doesn't have. All the function values between those two appear in both formulas. You can get a handle on this by comparing the three left rectangles from Figure 8-4 to the three right rectangles from Figure 8-6. Their areas and totals, which we calculated earlier, are

Three left rectangles: $1 + 2 + 5 = 8$

Three right rectangles: $2 + 5 + 10 = 17$

The sums of the areas are the same except for the left-most left rectangle and the right-most right rectangle. Both sums include the rectangles with areas 2 and 5. Look at how the rectangles are constructed — the second and third rectangles in Figure 8-4 are the same as the first and second rectangles in Figure 8-6.

Approximating area with midpoint sums

A third way to approximate areas with rectangles is to make each rectangle cross the curve at the midpoint of its top side. A midpoint sum is a *much* better estimate of area than either a left or a right sum. Figure 8-7 shows why.

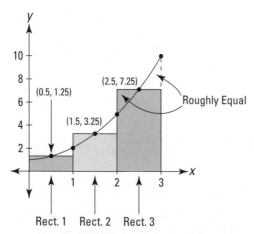

Figure 8-7: Three *midpoint* rectangles give you a much better estimate of the area under $f(x) = x^2 + 1$.

You can see in Figure 8-7 that the part of each rectangle that's above the curve looks about the same size as the gap between the rectangle and the curve. A midpoint sum produces such a good estimate because these two errors roughly cancel out each other.

For the three rectangles in Figure 8-8, the widths are 1 and the heights are $f(0.5) = 1.25$ $f(1.5) = 3.25$, and $f(2.5) = 7.25$. The total area comes to 11.75. Table 8-3 lists the midpoint sums for the same number of rectangles used in Tables 8-1 and 8-2.

Table 8-3 Estimates of the Area under $f(x) = x^2 + 1$ Given by Increasing Numbers of "Midpoint" Rectangles

Number of Rectangles	Area Estimate
3	11.75
6	11.9375
12	~11.9844
24	~11.9961
48	~11.9990
96	~11.9998
192	~11.9999
384	~11.99998

Sorry to give away the ending, but you'll soon see that the exact area is 12. And to see how much faster the midpoint approximations approach the exact answer of 12 than the left or right approximations, compare the three tables. The error with 6 midpoint rectangles is about the same as the error with 192 left or right rectangles!

The Midpoint Rule: You can approximate the exact area under a curve between a and b, $\int_a^b f(x)\,dx$, with a sum of *midpoint* rectangles given by the following formula. In general, the more rectangles, the better the estimate.

$$M_n = \frac{b-a}{n}\left[f\left(\frac{x_0+x_1}{2}\right)+f\left(\frac{x_1+x_2}{2}\right)+f\left(\frac{x_2+x_3}{2}\right)+\ldots\ldots\ldots \right.$$
$$\left. +f\left(\frac{x_{n-1}+x_n}{2}\right)\right]$$

where n is the number of rectangles, $\frac{b-a}{n}$ is the width of each, and the function values are the heights of the rectangles.

The left, right, and midpoint sums are called *Riemann sums* after the great German mathematician G. F. B. Riemann (1826–66).

Summation Notation

For adding up long series of numbers like the rectangle areas in a left, right, or midpoint sum, summation or *sigma* notation is handy.

Summing up the basics

Say you wanted to add up the first 100 multiples of 5 — that's from 5 to 500. You could write out the sum like this:

$$5 + 10 + 15 + 20 + 25 + \ldots\ldots\ldots + 490 + 495 + 500$$

But with sigma notation, the sum is much more condensed.

$$\sum_{i=1}^{100} 5i$$

This notation just tells you to plug 1 in for the i in $5i$, then plug 2 into the i in $5i$, then 3, then 4, all the way up to 100. Then you add up the results. So that's 5×1 plus 5×2 plus 5×3, and so on, up to 5×100. It's the same thing as writing out the sum the long way. By the way, the letter i has no significance. You can write the sum with a j, $\sum_{j=1}^{100} 5j$, or any other letter you like.

Here's one more. If you want to add up $10^2 + 11^2 + 12^2 + \ldots\ldots\ldots + 29^2 + 30^2$, you can write the sum with sigma notation as follows:

$$\sum_{k=10}^{30} k^2$$

Writing Riemann sums with sigma notation

You can use sigma notation to write out the right-rectangle sum for the curve $x^2 + 1$ from the "Approximating Area" sections. Recall the formula for a right sum from the "Approximating area with right sums" section:

$$R_n = \frac{b-a}{n}\left[f(x_1) + f(x_2) + f(x_3) + \ldots\ldots + f(x_n)\right]$$

Here's the same formula written with sigma notation:

$$R_n = \sum_{i=1}^{n}\left[f(x_i) \cdot \left(\frac{b-a}{n}\right)\right]$$

Now work this out for the six right rectangles in Figure 8-8. You're figuring the area under $x^2 + 1$ between 0 and 3 with six rectangles, so the width of each, $\frac{b-a}{n}$, is $\frac{3-0}{6}$, or $\frac{3}{6}$, or $\frac{1}{2}$. So now you've got

$$R_6 = \sum_{i=1}^{6}\left[f(x_i) \cdot \frac{1}{2}\right]$$

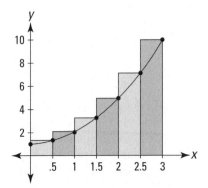

Figure 8-8: Six *right* rectangles approximate the area under $f(x) = x^2 + 1$ between 0 and 3.

Now, because the width of each rectangle is $\frac{1}{2}$, the right edges of the six rectangles fall on the first six multiples of $\frac{1}{2}$: 0.5, 1, 1.5, 2, 2.5, and 3. These numbers are the x-coordinates of the six points x_1 through x_6; they can be generated by the expression $\frac{1}{2}i$, where i equals 1 through 6. You can check that this works by plugging 1 in for i in $\frac{1}{2}i$, then 2, then 3, up to 6. So now you can replace the x_i in the formula with $\frac{1}{2}i$, giving you

$$R_6 = \sum_{i=1}^{6}\left[f\left(\frac{1}{2}i\right)\cdot\frac{1}{2}\right]$$

Our function, $f(x)$, is x^2+1 so $f\left(\frac{1}{2}i\right)=\left(\frac{1}{2}i\right)^2+1$, and so now you can write

$$R_6 = \sum_{i=1}^{6}\left[\left(\left(\frac{1}{2}i\right)^2+1\right)\cdot\frac{1}{2}\right]$$

If you plug 1 into i, then 2, 3, up to 6 and do the math, you get the sum of the areas of the rectangles in Figure 8-8. This sigma notation is just a fancy way of writing the sum of the six rectangles.

Are we having fun? Hold on, it gets worse. Now you're going to write out the general sum for an unknown number (n) of right rectangles. The total span of the area in question is 3, right? You divide this span by the number of rectangles to get the width of each rectangle. With 6 rectangles, the width of each is $\frac{3}{6}$; with n rectangles, the width of each is $\frac{3}{n}$. And the right edges of the n rectangles are generated by $\frac{3}{n}i$, for i equals 1 through n. That gives you

$$R_n = \sum_{i=1}^{n}\left[f\left(\frac{3}{n}i\right)\cdot\frac{3}{n}\right]$$

Or, because $f(x)=x^2+1$,

$$R_n = \sum_{i=1}^{n}\left[\left(\left(\frac{3}{n}i\right)^2 + 1\right)\cdot\frac{3}{n}\right]$$

$$= \sum_{i=1}^{n}\left[\left(\frac{9i^2}{n^2} + 1\right)\cdot\frac{3}{n}\right]$$

$$= \sum_{i=1}^{n}\left[\frac{27i^2}{n^3} + \frac{3}{n}\right]$$

$$= \sum_{i=1}^{n}\frac{27i^2}{n^3} + \sum_{i=1}^{n}\frac{3}{n} \quad \text{(take my word for it)}$$

$$= \frac{27}{n^3}\sum_{i=1}^{n}i^2 + \frac{3}{n}\sum_{i=1}^{n}1$$

For this last step, you pull the $\frac{27}{n^3}$ and the $\frac{3}{n}$ through the summation signs — you're allowed to pull out anything except for a function of i, the so-called *index of summation.* Also, the second summation in the last step has just a 1 after it and no i. So there's nowhere to plug in the values of i. No worries — all you do is add up n 1s, which equals n (I do this in a minute).

With sleight of hand, you're now going to turn the previous Riemann sum into a formula in terms of n. The sum of the first n square numbers, $1^2 + 2^2 + 3^2 + \ldots\ldots\ldots + n^2$, equals $\frac{n(n+1)(2n+1)}{6}$ (a formula you might want to memorize). So, you can substitute that expression for $\sum_{i=1}^{n}i^2$ in the last line of the sigma notation solution, and at the same time substitute n for $\sum_{i=1}^{n}1$:

$$R_n = \frac{27}{n^3}\cdot\frac{n(n+1)(2n+1)}{6} + \frac{3}{n}\cdot n$$

$$= \frac{27}{n^3}\cdot\left(\frac{n^3}{3} + \frac{n^2}{2} + \frac{n}{6}\right) + 3 \qquad \begin{array}{l}\text{(OK, I admit it, I didn't}\\\text{show all my work.)}\end{array}$$

$$= 9 + \frac{27}{2n} + \frac{9}{2n^2} + 3$$

$$= 12 + \frac{27}{2n} + \frac{9}{2n^2}$$

The end. Finally! This is the formula for the area of n right rectangles between 0 and 3 under the function $x^2 + 1$. You can use this formula to produce the results given in Table 8-2. But once you've got such a formula, it'd be kind of pointless to produce a table of approximate areas, because you can use the formula to determine the *exact* area. I get to that in the next section.

But first, here are the formulas for n left rectangles and n midpoint rectangles between 0 and 3 under the function $x^2 + 1$. These formulas generate the area approximations in Tables 8-1 and 8-3.

$$L_n = 12 - \frac{27}{2n} + \frac{9}{2n^2}$$

$$M_n = 12 - \frac{9}{4n^2}$$

Finding Exact Area with the Definite Integral

As you saw with the left, right, and midpoint rectangles in the "Approximating Area" sections, the more rectangles you have, the better the approximation. So, "all" you have to do to get the exact area under a curve is to use an infinite number of rectangles. Now, you can't really do that, but with the fantastic invention of limits, this is sort of what happens. Here's the definition of the definite integral that's used to compute exact areas.

The Definite Integral ("simple" definition): The exact area under a curve between a and b is given by the *definite integral*, which is defined as the limit of a Riemann sum:

$$\int_a^b f(x)\,dx = \lim_{n\to\infty} \sum_{i=1}^{n} \left[f(x_i) \cdot \left(\frac{b-a}{n} \right) \right]$$

Is that a thing of beauty or what? The summation above is identical to the formula for n right rectangles, R_n, from a few pages back. The only difference is that you take the limit of that formula as the number of rectangles approaches infinity.

This definition of the definite integral is a simple version based on the right rectangle formula. I'll skip the more complicated real-McCoy definition because you'll never need to use it. All Riemann sums have the same limit — it doesn't matter what type of rectangles you use — so you might as well use this right-rectangle definition.

Here, finally, is the exact area under $x^2 + 1$ between 0 and 3:

$$\int_0^3 \left(x^2 + 1\right) dx = \lim_{n \to \infty} \sum_{i=1}^{n} \left[f(x_i) \cdot \left(\frac{b-a}{n}\right) \right]$$

$$= \lim_{n \to \infty} \left(12 + \frac{27}{2n} + \frac{9}{2n^2} \right)$$ (This is what we got in the "Writing Riemann sums with sigma notation" section after all those steps.)

$$= 12 + \frac{27}{2 \cdot \infty} + \frac{9}{2 \cdot \infty^2}$$

$$= 12 + \frac{27}{\infty} + \frac{9}{\infty}$$

$$= 12 + 0 + 0$$ (Remember, in a limit problem, any number divided by infinity equals zero.)

$$= 12$$

Big surprise.

Chapter 9

Integration: Backwards Differentiation

Chapter 8 shows you the hard way to calculate the area under a curve using the formal definition of integration — the limit of a Riemann sum. In this chapter, I do it the easy way, taking advantage of one of the most important discoveries in mathematics — that integration is just differentiation in reverse.

Antidifferentiation: Reverse Differentiation

Antidifferentiation is just differentiation backwards. The derivative of $\sin x$ is $\cos x$, so the antiderivative of $\cos x$ is $\sin x$; the derivative of x^3 is $3x^2$, so the antiderivative of $3x^2$ is x^3 — you just go backwards . . . with one twist: The derivative of $x^3 + 10$ is also $3x^2$, as is the derivative of $x^3 - 5$. Any function of the form $x^3 + C$, where C is any number, has a derivative of $3x^2$. So, every such function is an antiderivative of $3x^2$.

The Indefinite Integral: The *indefinite integral* of a function $f(x)$, written as $\int f(x)\,dx$, is the family of *all* antiderivatives of the function. For example, because the derivative of x^3 is $3x^2$, the indefinite integral of $3x^2$ is $x^3 + C$, and you write

$$\int 3x^2 dx = x^3 + C$$

You may recognize this integration symbol, $\int$, from the *definite* integral in Chapter 8. The definite integral symbol, however, contains two little numbers like $\int_4^{10}$ that tell you to compute the area of a function between the two numbers, called the *limits of integration*. The naked version of the symbol, $\int$, indicates an *indefinite* integral or an *antiderivative*. This chapter is about the intimate connection between these two symbols.

Figure 9-1 shows the family of antiderivatives of the parabola $3x^2$, namely $x^3 + C$. Note that this family of curves has an infinite number of curves. They go up and down forever and are infinitely dense. The vertical gap of 2 units between each curve in Figure 9-1 is just a visual aid.

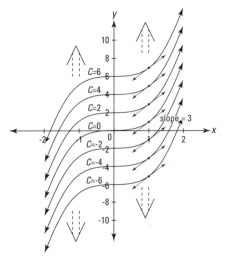

Figure 9-1: The family of curves $x^3 + C$. All these functions have the same derivative, $3x^2$.

The top curve on the graph is $x^3 + 6$; the one below it is $x^3 + 4$; the bottom one is $x^3 - 6$. By the power rule, these three functions, as well as all the others in this family of functions, have a derivative of $3x^2$. Now consider the slope of each of the curves where x equals 1 (see the tangent lines drawn on the curves). The derivative of each curve is $3x^2$, so when x equals 1, the slope of each curve is $3 \cdot 1^2$, or 3. Thus, all these little tangent lines are parallel. No matter what x-value you pick, tangent lines like this would be parallel. This is the visual way to understand why each of these curves has the same derivative, and, thus, why each curve is an antiderivative of the same function.

The Annoying Area Function

This topic is pretty tricky. Put on your thinking cap. Say you've got any old function, $f(t)$. Imagine that at some t-value, call it s, you draw a fixed vertical line. See Figure 9-2.

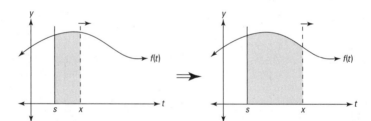

Figure 9-2: Area under f between s and x is swept out by the moving line at x.

Then you take a movable vertical line, starting at the same point, s (for *starting* point), and drag it to the right, sweeping out a larger and larger area under the curve. This area is a function of x, the position of the moving line. In symbols, you write

$$A_f(x) = \int_s^x f(t)\,dt$$

Note that t is the input variable in $f(t)$ instead of x because x is already taken — it's the input variable in $A_f(x)$. The subscript f in A_f indicates that $A_f(x)$ is the area function for the particular curve f or $f(t)$. The dt is a little increment along the t-axis — actually an infinitesimally small increment.

Here's a simple example to make sure you've got a handle how an area function works. Say you've got the simple function, $f(t) = 10$ — that's a horizontal line at $y = 10$. If you sweep out area beginning at $s = 3$, you get the following area function:

$$A_f(x) = \int_3^x 10\,dt$$

You can see that the area swept out from 3 to 4 is 10 because, in dragging the line from 3 to 4, you sweep out a rectangle with a width of 1 and a height of 10, which has an area of 1 times 10, or 10. See Figure 9-3.

So, $A_f(4)$, the area swept out as you hit 4, equals 10. $A_f(5)$ equals 20 because when you drag the line to 5, you've swept out a rectangle with a width of 2 and height of 10, which has an area of 2 times 10, or 20. $A_f(6)$ equals 30, and so on.

Now, imagine that you drag the line across at a rate of one *unit per second.* You start at $x = 3$, and you hit 4 at 1 second, 5 at 2 seconds, 6 at 3 seconds, and so on. How much area are you sweeping out per second? Ten *square units per second* because each second you sweep out another 1-by-10 rectangle. Notice — this is huge — that because the width of each rectangle you sweep out is 1, the area of each rectangle — given by *height* times *width* — is the same as its height because anything times 1 equals itself.

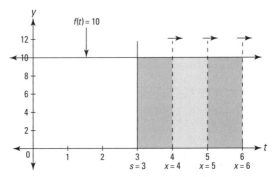

Figure 9-3: Area under $f = 10$ between 3 and x is swept out by the moving vertical line at x.

Okay, are you sitting down? You've reached another one of the big *Ah ha!* moments in the history of mathematics. Recall that a derivative is a rate. So, because the rate at which the previous area function grows is 10 *square units per second,* you can say its derivative equals 10. Thus, you can write

$$\frac{d}{dx} A_f(x) = 10$$

Here's the critical thing: This rate or derivative of 10 is the same as the original function $f(t) = 10$ because as you go across 1 unit, you sweep out a rectangle that has an area of 10, the height of the function.

This works for any function, not just horizontal lines. Look at the function $g(t)$ and its area function $A_g(x)$ that sweeps out area beginning at $s = 2$ in Figure 9-4.

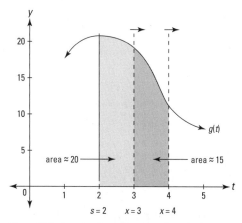

Figure 9-4: Area under *g* between 2 and *x* is swept out by the moving vertical line at *x*.

You can see that $A_g(3)$ is about 20 because the area swept out between 2 and 3 has a width of 1 and the curved top of the "rectangle" has an average height of about 20. So, during this interval, the rate of growth of $A_g(x)$ is about 20 *square units per second.* Between 3 and 4, you sweep out about 15 square units of area because that's roughly the average height of $g(t)$ between 3 and 4. So, during second number two — the interval from $x = 3$ to $x = 4$ — the rate of growth of $A_g(x)$ is about 15.

The *rate* of area being swept out under a curve by an area function at a given x-value is equal to the *height* of the curve at that x-value.

The Fundamental Theorem

Sound the trumpets! Now that you've seen the connection between the rate of growth of an area function and the height of the given curve, you're ready for what some say is one of the most important theorems in the history of mathematics:

The Fundamental Theorem of Calculus: Given an area function A_f that sweeps out area under $f(t)$,

$$A_f(x) = \int_s^x f(t)\,dt,$$

the rate at which area is being swept out is equal to the height of the original function. So, because the rate is the derivative, the derivative of the area function equals the original function:

$$\frac{d}{dx}A_f(x) = f(x).$$

Because $A_f(x) = \int_s^x f(t)\,dt$, you can also write the above equation as follows:

$$\frac{d}{dx}\int_s^x f(t)\,dt = f(x)$$

Now, because the derivative of $A_f(x)$ is $f(x)$, $A_f(x)$ is by definition an antiderivative of $f(x)$. Check out how this works by returning to the simple function from the previous section, $f(t) = 10$, and its area function, $A_f(x) = \int_s^x 10\,dt$.

According to the Fundamental Theorem, $\frac{d}{dx}A_f(x) = 10$. Thus A_f must be an antiderivative of 10; in other words, A_f is a function whose derivative is 10. Because any function of the form $10x + C$, where C is a number, has a derivative of 10, the antiderivative of 10 is $10x + C$. The particular number C depends on your choice of s, the point where you start sweeping out area. For a particular choice of s, the area function will be the one function (out of all the functions in the family of curves $10x + C$) that crosses the x-axis at s. To figure out C, set the antiderivative equal to zero, plug the value of s into x, and solve for C.

For this function with an antiderivative of $10x + C$, if you start sweeping out area at, say, $s = 0$, then $10 \cdot 0 + C = 0$, so $C = 0$, and thus $A_f(x) = \int_0^x 10\,dt = 10x + 0$, or just $10x$. If instead you start sweeping out area at $s = -2$ and define a new area function, $B_f(x) = \int_{-2}^x 10\,dt$, then $10 \cdot (-2) + C = 0$, so C equals 20 and $B_f(x)$ is thus $10x + 20$. This area function is 20 more than $A_f(x)$, which starts at $s = 0$, because if you start at $s = -2$, you've already swept out an area of 20 by the time you get to zero.

And if you start sweeping out area at $s = 3$, $10 \cdot 3 + C = 0$, so $C = -30$ and the area function is $C_f(x) = \int_3^x 10\,dt = 10x - 30$. This function is 30 less than on $A_f(x)$ because with $C_f(x)$, you lose the 3-by-10 rectangle between 0 and 3 that $A_f(x)$ has.

The area swept out under the horizontal line $f(t) = 10$, from some number s to x, is given by an antiderivative of 10, namely $10x + C$, where the value of C depends on where you start sweeping out area.

For the next example, look again at the parabola $x^2 + 1$, our friend from Chapter 8, and the discussion of Riemann sums. Flip back to Figure 8-5. Now you can finally compute the exact area (from 0 to 3) in that graph the easy way.

The area function for sweeping out area under $x^2 + 1$ is $A_f(x) = \int_s^x (t^2 + 1)\,dt$. By the Fundamental Theorem, $\frac{d}{dx} A_f(x) = x^2 + 1$, and so A_f is an antiderivative of $x^2 + 1$. Any function of the form $\frac{1}{3}x^3 + x + C$ has a derivative of $x^2 + 1$ (try it), so that's the antiderivative. For Figure 8-5, you want to sweep out area beginning at 0, so $s = 0$. Plug 0 into the antiderivative and solve for C: $\frac{1}{3}0^3 + 0 + C = 0$, so $C = 0$, and thus

$$A_f(x) = \int_0^x (t^2 + 1)dt = \frac{1}{3}x^3 + x$$

The area swept out from 0 to 3 — which we did the hard way in Chapter 8 by computing the limit of a Riemann sum — is simply $A_f(3)$:

$$A_f(x) = \frac{1}{3}x^3 + x$$
$$A_f(3) = \frac{1}{3} \cdot 3^3 + 3$$
$$= 9 + 3$$
$$= 12$$

That was *much* less work than doing it the hard way.

And after you know the area function that starts at zero, $\int_0^x (t^2 + 1)dt = \frac{1}{3}x^3 + x$, it's a snap to figure the area of other sections under the parabola that don't start at zero. Say you want the area under the parabola between 2 and 3. You can compute it by subtracting the area between 0 and 2 from the area between 0 and 3. You just figured the area between 0 and 3 — that's 12. And the area between 0 and 2 is $A_f(2) = \frac{1}{3} \cdot 2^3 + 2 = 4\frac{2}{3}$. So the area between 2 and 3 is $12 - 4\frac{2}{3}$, or $7\frac{1}{3}$. This subtraction method brings us to the next topic — the second version of the Fundamental Theorem.

Fundamental Theorem: Take Two

Now we finally arrive at the super-duper shortcut integration theorem. But first a warning. . . .

When using an area function, the first version of the Fundamental Theorem of Calculus, or its second version, areas *below* the x-axis count as *negative* areas.

The Fundamental Theorem of Calculus (shortcut version): Let F be any antiderivative of the function f; then

$$\int_a^b f(x)dx = F(b) - F(a)$$

This theorem gives you the super shortcut for computing a definite integral like $\int_{2}^{3}(x^2+1)dx$, the area under the parabola x^2+1 between 2 and 3. As I show in the previous section, you can get this area by subtracting the area between 0 and 2 from the area between 0 and 3, but to do that you need to know that the particular area function sweeping out area beginning at zero, $\int_{0}^{x}(t^2+1)dt$, is $\frac{1}{3}x^3+x$ (with a C value of zero).

The beauty of the shortcut theorem is that you don't have to even use an area function like $A_f(x)=\int_{0}^{x}(t^2+1)dt$. You just find any antiderivative, $F(x)$, of your function, and do the subtraction, $F(b)-F(a)$. The simplest antiderivative to use is the one where $C=0$. So here's how you use the theorem to find the area under our parabola from 2 to 3. $F(x)=\frac{1}{3}x^3+x$ is an antiderivative of x^2+1 so, by the theorem,

$$\int_{2}^{3}(x^2+1)dx = F(3)-F(2)$$

$F(3)-F(2)$ can be written as $\left[\frac{1}{3}x^3+x\right]_{2}^{3}$, and thus,

$$\int_{2}^{3}(x^2+1)dx = \left[\frac{1}{3}x^3+x\right]_{2}^{3}$$
$$= \frac{1}{3}\cdot 3^3 + 3 - \left(\frac{1}{3}\cdot 2^3 + 2\right)$$
$$= 12 - 4\tfrac{2}{3}$$
$$= 7\tfrac{1}{3}$$

Granted, this is the same computation I did in the previous section using the area function with $s=0$, but that's only because for the x^2+1 function, when s is zero, C is also zero. It's sort of a coincidence, and it's not true for all functions. But regardless of the function, the shortcut works, and you don't have to worry about area functions or s or C. All you do is $F(b)-F(a)$.

Here's another example: What's the area under $f(x)=e^x$ between $x=3$ and $x=5$? The derivative of e^x is e^x, so e^x is an antiderivative of e^x, and thus

$$\int_{3}^{5} e^x \, dx = \left[e^x \right]_{3}^{5}$$

$$= e^5 - e^3$$

$$\approx 148.4 - 20.1$$

$$\approx 128.3$$

What could be simpler? And if one big shortcut wasn't enough to make your day, Table 9-1 lists some rules about definite integrals that can make your life much easier.

Table 9-1	Five Easy Rules for Definite Integrals
1)	$\int_{a}^{a} f(x)dx = 0$ (Well, duh, there's no "area" between a and a)
2)	$\int_{b}^{a} f(x)dx = -\int_{a}^{b} f(x)dx$
3)	$\int_{a}^{b} f(x)dx = \int_{a}^{c} f(x)dx + \int_{c}^{b} f(x)dx$
4)	$\int_{a}^{b} k\, f(x)dx = k\int_{a}^{b} f(x)dx$ (k is a constant; you can pull a constant out of the integral
5)	$\int_{a}^{b} \left[f(x) \pm g(x) \right]dx = \int_{a}^{b} f(x)dx \pm \int_{a}^{b} g(x)dx$

Antiderivatives: Basic Techniques

This section gives some basic techniques for antiderivatives.

Reverse rules

The easiest antiderivatives are ones that are the reverse of derivative rules you already know. These are automatic, one-step antiderivatives with the exception of the reverse power rule, which is only slightly harder.

No-brainer reverse rules

You know that the derivative of sin x is cos x, so reversing that tells you that an antiderivative of cos x is sin x. What could be simpler? But don't forget that all functions of the form sin $x + C$ are antiderivatives of cos x. In symbols, you write

$$\frac{d}{dx}\sin x = \cos x, \text{ and therefore}$$

$$\int \cos x = \sin x + C$$

Table 9-2 lists the reverse rules for antiderivatives.

Table 9-2	Basic Antiderivative Formulas
1) $\int 1dx$ (or just $\int dx$) $= x + C$ (because the derivative of x is 1)	
2) $\int x^n dx = \frac{x^{n+1}}{n+1} + C \ (n \neq -1)$	3) $\int \frac{dx}{x} = \ln|x| + C$
4) $\int e^x dx = e^x + C$	5) $\int a^x dx = \frac{1}{\ln a} a^x + C$
6) $\int \sin x \ dx = -\cos x + C$	7) $\int \cos x \ dx = \sin x + C$
8) $\int \sec^2 x \ dx = \tan x + C$	9) $\int \csc^2 x \ dx = -\cot x + C$
10) $\int \sec x \tan x \ dx = \sec x + C$	11) $\int \csc x \cot x \ dx = -\csc x + C$
12) $\int \frac{dx}{\sqrt{a^2 - x^2}} = \arcsin \frac{x}{a} + C$	13) $\int \frac{dx}{a^2 + x^2} = \frac{1}{a} \arctan \frac{x}{a} + C$
14) $\int \frac{dx}{x\sqrt{x^2 - a^2}} = \frac{1}{a} \text{arcsec} \frac{|x|}{a} + C$	

The slightly more difficult reverse power rule

By the power rule, you know that

$$\frac{d}{dx} x^3 = 3x^2, \text{ and therefore}$$

$$\int 3x^2 \ dx = x^3 + C$$

Here's the simple method for reversing the power rule. Use $5x^4$ for your function. Recall that the power rule says to

1. **Bring the power in front where it will *multiply* the rest of the derivative.**

   $$5x^4 \rightarrow 4 \cdot 5x^4$$

2. ***Reduce* the power by one and simplify.**

   $$4 \cdot 5x^4 \rightarrow 4 \cdot 5x^3 = 20x^3$$

To reverse this process, reverse the order of the two steps and reverse the math within each step. Here's how it works:

1. ***Increase* the power by one.**

   The 3 becomes a 4.

   $$20x^3 \rightarrow 20x^4$$

2. ***Divide* by the new power and simplify.**

   $$20x^4 \rightarrow \frac{20}{4}x^4 = 5x^4$$

And thus you write $\int 20x^3\, dx = 5x^4 + C$.

Especially when you're new to antidifferentiation, it's a good idea to test your antiderivatives by differentiating them — you can ignore the C. If you get back to your original function, you know your antiderivative is correct.

With the antiderivative you just found and the second version of the Fundamental Theorem, you can determine the area under $20x^3$ between, say, 1 and 2:

$$\int 20x^3\, dx = 5x^4 + C, \text{ thus}$$

$$\int_1^2 20x^3\, dx = \left[5x^4 \right]_1^2$$

$$= 5 \cdot 2^4 - 5 \cdot 1^4$$

$$= 80 - 5$$

$$= 75$$

Guess and check

The guess-and-check method works when the *integrand* (that's the expression after the integral symbol not counting the *dx*, and it's the thing you want to antidifferentiate) is close to a function that you know the reverse rule for. For example, say you want the antiderivative of $\cos(2x)$. Well, you know that the derivative of sine is cosine. Reversing that tells you that the antiderivative of cosine is sine. So you might think that the antiderivative of $\cos(2x)$ is $\sin(2x)$. That's your *guess*. Now *check* it by differentiating it to see if you get the original function, $\cos(2x)$:

$$\frac{d}{dx}\sin(2x)$$
$$= \cos(2x) \cdot 2 \ (\text{sine rule and chain rule})$$
$$= 2\cos(2x)$$

This result is very close to the original function, except for that extra coefficient of 2. In other words, the answer is 2 times as much as what you want. Because you want half that result, try an antiderivative that's half of your first guess: $\frac{1}{2}\sin(2x)$. Check this second guess by differentiating it, and you get the desired result.

Here's another example. What's the antiderivative of $(3x-2)^4$?

1. **Guess the antiderivative.**

   This looks sort of like a power rule problem, so try the reverse power rule. The antiderivative of x^4 is $\frac{1}{5}x^5$ by the reverse power rule, so your guess is $\frac{1}{5}(3x-2)^5$.

2. **Check your guess by differentiating it.**

   $$\frac{d}{dx}\left[\frac{1}{5}(3x-2)^5\right]$$
   $$= 5 \cdot \frac{1}{5}(3x-2)^4 \cdot 3 \ (\text{power rule and chain rule})$$
   $$= 3(3x-2)^4$$

3. Tweak your first guess.

Your result, $3(3x-2)^4$, is three times too much, so make your second guess a *third* of your first guess — that's $\frac{1}{3} \cdot \frac{1}{5}(3x-2)^5$, or $\frac{1}{15}(3x-2)^5$.

4. Check your second guess by differentiating it.

$$\frac{d}{dx}\left[\frac{1}{15}(3x-2)^5\right]$$

$$= 5 \cdot \frac{1}{15}(3x-2)^4 \cdot 3 \;(\text{power rule and chain rule})$$

$$= (3x-2)^4$$

This checks. You're done. The antiderivative of $(3x-2)^4$ is $\frac{1}{15}(3x-2)^5 + C$

The two previous examples show that *guess and check* works well when the function you want to antidifferentiate has an argument like $3x$ or $3x + 2$ (where x is raised to the *first* power) instead of a plain old x. (Recall that in a function like $\sqrt{5x}$, the $5x$ is called the *argument.*) In this case, you just tweak your guess by the *reciprocal* of the coefficient of x — the 3 in $3x + 2$, for example (the 2 in $3x + 2$ has no effect on your answer). In fact, for these easy problems, you don't really have to guess and check. You can immediately see how to tweak your guess. It becomes sort of a one-step process. If the function's argument is more complicated than $3x + 2$ — like the x^2 in $\cos(x^2)$ — you have to try the next method, substitution.

Substitution

In the previous section, you can see why the first guess in each case didn't work. When you differentiate the guess, the chain rule produces an extra constant: 2 in the first example, 3 in the second. You then tweak the guesses with 1/2 and 1/3 to compensate for the extra constant.

Now say you want the antiderivative of $\cos(x^2)$ and you guess that it is $\sin(x^2)$. Watch what happens when you differentiate $\sin(x^2)$ to check it.

$$\frac{d}{dx}\sin\left(x^2\right)$$

$$= \cos\left(x^2\right)\cdot 2x \ \left(\text{sine rule and chain rule}\right)$$

$$= 2x\cos\left(x^2\right)$$

Here the chain rule produces an extra $2x$ — because the derivative of x^2 is $2x$ — but if you try to compensate for this by attaching a $\frac{1}{2x}$ to your guess, it won't work. Try it.

So, guessing and checking doesn't work for antidifferentiating $\cos\left(x^2\right)$ — actually *no* method works for this simple-looking integrand (not all functions have antiderivatives) — but your attempt at differentiation here reveals a new class of functions that you can antidifferentiate. Because the derivative of $\sin\left(x^2\right)$ is $2x\cos\left(x^2\right)$, the antiderivative of $2x\cos\left(x^2\right)$ must be $\sin\left(x^2\right)$. This function, $2x\cos\left(x^2\right)$, is the type of function you antidifferentiate with the substitution method.

The substitution method works when the integrand contains a function and *the derivative of the function's argument* — in other words, when it contains that extra thing produced by the chain rule — or something just like it except for a constant. And the integrand must not contain anything else.

The derivative of e^{x^3} is $e^{x^3}\cdot 3x^2$ by the e^x rule and the chain rule. So, the antiderivative of $e^{x^3}\cdot 3x^2$ is e^{x^3}. And if you were asked to find the antiderivative of $e^{x^3}\cdot 3x^2$, you would know that the substitution method would work because this expression contains $3x^2$, which is the derivative of the argument of e^{x^3}, namely x^3.

You may be wondering why this is called the substitution method. I show you why in a minute. But first, I want to point out that you don't always have to use the step-by-step method. Assuming you understand why the antiderivative of $e^{x^3}\cdot 3x^2$ is e^{x^3}, you may encounter problems where you can just see the antiderivative without doing any work. But whether or not you can just see the answers to problems like this one, the substitution method is a good one to learn because, for one thing, it has many uses in calculus and other areas of mathematics, and for another, your teacher may require that you know it and use it. So here's how to find the antiderivative of $\int 2x\cos\left(x^2\right)dx$ with substitution.

1. **Set u equal to the argument of the main function.**

   The argument of $\cos\left(x^2\right)$ is x^2, so you set u equal to x^2.

2. **Take the derivative of u with respect to x.**

$$u = x^2 \text{ so } \frac{du}{dx} = 2x$$

3. **Solve for dx.**

$$\frac{du}{dx} = \frac{2x}{1}$$

$$du = 2x\,dx \ \left(\text{cross multiplication}\right)$$

$$\frac{du}{2x} = dx \ \left(\text{dividing both sides by } 2x\right)$$

4. **Make the substitutions.**

   In $\int 2x\cos\left(x^2\right)dx$, u takes the place of x^2 and $\frac{du}{2x}$ takes the place of dx. So now you've got $\int 2x\cos u\frac{du}{2x}$. The two $2x$s cancel, giving you $\int \cos u\,du$.

5. **Antidifferentiate using the simple reverse rule.**

$$\int \cos u\,du = \sin u + C$$

6. **Substitute x^2 back in for u — coming full circle.**

   u equals x^2, so x^2 goes in for the u:

$$\int \cos u\,du = \sin\left(x^2\right) + C$$

   That's it. So $\int 2x\cos\left(x^2\right)dx = \sin\left(x^2\right) + C$.

If the original problem had been $\int 5x\cos\left(x^2\right)dx$ instead of $\int 2x\cos\left(x^2\right)dx$, you follow the same steps except that in Step 4, after making the substitution, you arrive at $\int 5x\cos u\frac{du}{2x}$. The xs still cancel — that's the important thing — but after canceling you get $\int \frac{5}{2}\cos u\,du$, which has that extra $\frac{5}{2}$ in it. No worries. Just pull the $\frac{5}{2}$ through the $\int$ giving you $\frac{5}{2}\int \cos u\,du$. Now you finish this problem just as you did above in Steps 5 and 6, except for the extra $\frac{5}{2}$.

$$\frac{5}{2}\int \cos u\, du = \frac{5}{2}\left(\sin u + C\right)$$

$$= \frac{5}{2}\sin u + \frac{5}{2}C$$

$$= \frac{5}{2}\sin\left(x^2\right) + \frac{5}{2}C$$

Because C is any old constant, $\frac{5}{2}C$ is still any old constant, so you can get rid of the $\frac{5}{2}$ in front of the C. That may seem somewhat unmathematical, but it's right. Thus, your final answer is $\frac{5}{2}\sin\left(x^2\right) + C$. You should check this by differentiating it.

Here are a few examples of antiderivatives you can do with the substitution method so you can learn how to spot them.

✔ $\int 4x^2 \cos\left(x^3\right) dx$

The derivative of x^3 is $3x^2$, but you don't have to pay any attention to the 3 in $3x^2$ or the 4 in the integrand. Because the integrand contains x^2, and because it doesn't contain any other extra stuff, substitution works. Try it.

✔ $\int 10\sec^2 x \cdot e^{\tan x}\, dx$

The integrand contains a function, $e^{\tan x}$, as well as the derivative of its argument, $\tan x$ — which is $\sec^2 x$. Because the integrand doesn't contain any other extra stuff (except for the 10, which doesn't matter), substitution works. Do it.

✔ $\int \frac{2}{3}\cos x \sqrt{\sin x}\; dx$

Because the integrand contains the derivative of $\sin x$, namely $\cos x$, and no other stuff except for the 2/3, substitution works. Go for it.

You can do the three problems just listed with a method that combines substitution and guess and check (as long as your teacher doesn't insist that you show the six-step substitution solution). Try using this combo method to antidifferentiate the first example, $\int 4x^2 \cos\left(x^3\right) dx$. First confirm that the integral fits the pattern for substitution — it does, as seen in the first checklist item. This confirmation is the only part substitution plays in the combo method. Now finish the problem with guess and check.

1. **Make your guess.**

   The antiderivative of cosine is sine, so a good guess for the antiderivative of $4x^2 \cos(x^3)$ is $\sin(x^3)$.

2. **Check your guess by differentiating it.**

   $$\frac{d}{dx}\sin(x^3) = \cos(x^3) \cdot 3x^2 \quad (\text{sine rule and chain rule})$$
   $$= 3x^2 \cos(x^3)$$

3. **Tweak your guess.**

   Your result from Step 2, $3x^2 \cos(x^3)$, is 3/4 of what you want, $4x^2 \cos(x^3)$, so make your guess 4/3 bigger (note that 4/3 is the reciprocal of 3/4). Your second guess is thus $\frac{4}{3}\sin(x^3)$.

4. **Check this second guess by differentiating it.**

   Oh, heck, skip this — your answer's got to work.

Chapter 10

Integration for Experts

• •

In This Chapter

▶ Breaking down integrals into parts

▶ Finding trigonometric integrals

▶ Understanding the *A*s, *B*s, and *C*s of partial fractions

• •

I figure it wouldn't hurt to give you a break from the kind of theoretical groundwork stuff that I lay on pretty thick in Chapter 9, so this chapter cuts to the chase and shows you just the nuts and bolts of several integration techniques.

Integration by Parts

The basic idea of integration by parts is to transform an integral you *can't* do into a simple product minus an integral you *can* do. Here's the formula:

Integration by Parts: $\int u\,dv = uv - \int v\,du$

And here's the method in a nutshell. What's $\int \sqrt{x}\ln(x)dx$? First, you've got to split up the integrand into a *u* and a *dv* so that it fits the formula. For this problem, choose $\ln(x)$ to be your *u*. Then everything else is the *dv*, namely $\sqrt{x}\ dx$. (I show you how to choose *u* in the next section.) Next, you differentiate *u* to get your *du*, and you integrate *dv* to get your *v*. Finally, you plug everything into the formula and you're home free.

To help keep everything straight, organize integration-by-parts problems with a box like the one on the left in Figure 10-1. For each new problem, draw an empty 2-by-2 box, then put your *u* in the upper-left corner and your *dv* in the lower-right corner. The box for the above problem is on the right in Figure 10-1.

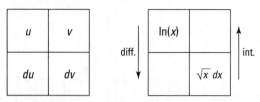

Figure 10-1: The integration-by-parts box.

The arrows in Figure 10-1 remind you to differentiate on the left and to integrate on the right. Now do the calculus and complete the box as shown in Figure 10-2:

$$u = \ln(x) \quad dv = \sqrt{x}\ dx$$
$$\frac{du}{dx} = \frac{1}{x} \quad \int dv = \int \sqrt{x}\ dx$$
$$du = \frac{1}{x}dx \quad v = \frac{2}{3}x^{3/2}$$

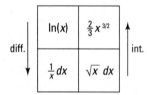

Figure 10-2: The completed box for $\int \sqrt{x} \ln(x) dx$.

Ready to finish? Plug everything into the formula:

$$\int u\, dv = uv - \int v\, du$$
$$\int \sqrt{x} \ln(x) dx = \ln(x) \cdot \frac{2}{3}x^{3/2} - \int \frac{2}{3}x^{3/2} \cdot \frac{1}{x}dx$$
$$= \frac{2}{3}x^{3/2} \ln(x) - \frac{2}{3}\int x^{1/2}dx$$
$$= \frac{2}{3}x^{3/2} \ln(x) - \frac{2}{3}\left(\frac{2}{3}x^{3/2} + C\right) \quad \text{(reverse power rule)}$$
$$= \frac{2}{3}x^{3/2} \ln(x) - \frac{4}{9}x^{3/2} - \frac{2}{3}C$$
$$= \frac{2}{3}x^{3/2} \ln(x) - \frac{4}{9}x^{3/2} - C$$

In the last step, you replace the $-\frac{2}{3}C$ with C because $-\frac{2}{3}$ times any old number is still just any old number.

Picking your u

Here's a great mnemonic for how to choose the *u* (again, once you've picked your *u*, everything else is the *dv*).

Herbert E. Kasube came up with the acronym *LIATE* to help you choose your *u* (calculus nerds can check out Herb's article in the *American Mathematical Monthly* 90, 1983 issue):

L	Logarithmic	(like log(x))
I	Inverse trigonometric	(like arctan(x))
A	Algebraic	(like $5x^2 + 3$)
T	Trigonometric	(like cos(x))
E	Exponential	(like 10^x)

To pick your *u*, go down this list; the first type of function on it that appears in the integrand is the *u*. Example: Integrate $\int \arctan(x)dx$. (Integration by parts sometimes works for integrands like this one that contain only a single function.)

1. **Go down the LIATE list and pick the *u*.**

   You see that there are no logarithmic functions in $\arctan(x)dx$, but there is an inverse trigonometric function, $\arctan(x)$. So that's your *u*. Everything else is your *dv*, namely, plain old *dx*.

2. **Do the box thing.** See Figure 10-3.

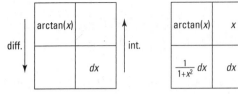

Figure 10-3: The box thing.

3. **Plug everything into the integration-by-parts formula.**

$$\int u\,dv = uv - \int v\,du$$

$$\int \arctan(x)\,dx = x\arctan(x) - \int x \cdot \frac{1}{1+x^2}\,dx$$

Now you can finish this problem by integrating $\int x \cdot \frac{1}{1+x^2}\,dx$ with the substitution method, setting $u = 1+x^2$. Try it. Note that the u in $u = 1+x^2$ has nothing to do with integration-by-parts u. Your final answer should be $\int \arctan(x)\,dx = x\arctan(x) - \frac{1}{2}\ln(1+x^2) + C$.

Here's another one. Integrate $\int x\sin(3x)\,dx$.

1. **Go down the *LIATE* list and pick the *u*.**

   Going down the *LIATE* list, the first type of function you find in $x\sin(3x)\,dx$ is a very simple algebraic one, namely x, so that's your u.

2. **Do the box thing.** See Figure 10-4.

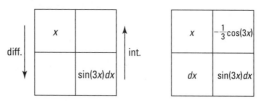

Figure 10-4: Yet more boxes.

3. **Plug everything into the integration-by-parts formula.**

$$\int u\,dv = uv - \int v\,du$$

$$\int x\sin(3x)\,dx = -\frac{1}{3}x\cos(3x) - \int -\frac{1}{3}\cos(3x)\,dx$$

$$= -\frac{1}{3}x\cos(3x) + \frac{1}{3}\int \cos(3x)\,dx$$

You can easily integrate $\int \cos(3x)\,dx$ with substitution or the guess-and-check method. Your final answer should be $-\frac{1}{3}x\cos(3x) + \frac{1}{9}\sin(3x) + C$.

Tricky Trig Integrals

In this section, you integrate powers of the six trigonometric functions, like $\int \sin^3(x)\,dx$ and $\int \sec^4(x)\,dx$, and products or quotients of trig functions, like $\int \sin^2(x)\cos^3(x)\,dx$ and $\int \dfrac{\csc^2(x)}{\cot(x)}\,dx$. This is a bit tedious — time for some caffeine.

To use the following techniques, you must have an integrand that contains just one of the six trig functions like $\int \csc^3(x)\,dx$ or a certain pairing of trig functions, like $\int \sin^2(x)\cos(x)\,dx$. If the integrand has two trig functions, the two must be one of these three pairs: sine with cosine, secant with tangent, or cosecant with cotangent. For an integrand containing something other than one of these pairs, you can convert the problem into one of these pairs by using trig identities like $\sin(x) = 1/\csc(x)$ and $\tan(x) = \sin(x)/\cos(x)$. For instance,

$$\int \sin^2(x)\sec(x)\tan(x)\,dx$$

$$= \int \sin^2(x)\frac{1}{\cos(x)} \cdot \frac{\sin(x)}{\cos(x)}\ dx$$

$$= \int \frac{\sin^3(x)}{\cos^2(x)}\ dx$$

After any needed conversions, you get one of three cases:

$$\int \sin^m(x)\cos^n(x)\,dx$$

$$\int \sec^m(x)\tan^n(x)\,dx$$

$$\int \csc^m(x)\cot^n(x)\,dx$$

where either m or n is a positive integer.

Positive powers of trig functions are generally preferable to negative powers, so, for example, you want to convert $\int \sin^{-2}(x)\tan^{-2}(x)\,dx$ into $\int \csc^2(x)\cot^2(x)\,dx$.

Sines and cosines

This section covers integrals containing sines and cosines.

Case 1: The power of sine is odd and positive

If the power of sine is odd and positive, lop off one sine factor and put it to the right of the rest of the expression, convert the remaining sine factors to cosines with the Pythagorean identity, and then integrate with the substitution method where $u = \cos(x)$.

The *Pythagorean identity* tells you that, for any angle x, $\sin^2(x) + \cos^2(x) = 1$. And thus $\sin^2(x) = 1 - \cos^2(x)$ and $\cos^2(x) = 1 - \sin^2(x)$.

Now integrate $\int \sin^3(x)\cos^4(x)dx$.

1. **Lop off one sine factor and move it to the right.**

$$\int \sin^3(x)\cos^4(x)dx = \int \sin^2(x)\cos^4(x)\sin(x)dx$$

2. **Convert the remaining sines to cosines using the Pythagorean identity and simplify.**

$$\int \sin^2(x)\cos^4(x)\sin(x)dx$$
$$= \int \left(1 - \cos^2(x)\right)\cos^4(x)\sin(x)dx$$
$$= \int \left(\cos^4(x) - \cos^6(x)\right)\sin(x)dx$$

3. **Integrate with substitution, where $u = \cos(x)$.**

$$u = \cos(x)$$
$$\frac{du}{dx} = -\sin(x)$$
$$du = -\sin(x)dx$$

You can save a little time in all substitution problems by just solving for *du* — as I did immediately above — and not bothering to solve for *dx*. You then tweak the integral so that it contains the thing *du* equals ($-\sin(x)dx$) in this problem. The integral contains a $\sin(x)dx$, so you multiply it by -1 to turn it into $-\sin(x)dx$ and then compensate for that -1 by multiplying the whole integral by -1. This is a wash because

−1 times −1 equals 1. This may not sound like much of a shortcut, but it's a good time-saver once you get used to it.

So tweak your integral:

$$\int \left(\cos^4(x) - \cos^6(x)\right)\left(\sin(x)dx\right)$$
$$= -\int \left(\cos^4(x) - \cos^6(x)\right)\left(-\sin(x)dx\right)$$

Now substitute and solve by the reverse power rule:

$$= -\int \left(u^4 - u^6\right)du$$
$$= -\frac{1}{5}u^5 + \frac{1}{7}u^7 + C$$
$$= \frac{1}{7}\cos^7(x) - \frac{1}{5}\cos^5(x) + C$$

Case 2: The power of cosine is odd and positive

This problem works exactly like Case 1, except that the roles of sine and cosine are reversed. Find $\int \dfrac{\cos^3(x)}{\sqrt{\sin(x)}}\,dx$.

1. Lop off one cosine factor and move it to the right.

$$\int \frac{\cos^3(x)}{\sqrt{\sin(x)}}\,dx = \int \cos^3(x)\left(\sin^{-1/2}(x)\right)dx$$
$$= \int \cos^2(x)\left(\sin^{-1/2}(x)\right)\cos(x)dx$$

2. Convert the remaining cosines to sines with the Pythagorean identity and simplify.

$$= \int \left(1 - \sin^2(x)\right)\left(\sin^{-1/2}(x)\right)\cos(x)dx$$
$$= \int \left(\sin^{-1/2}(x) - \sin^{3/2}(x)\right)\cos(x)dx$$

3. Integrate with substitution, where $u = \sin(x)$.

$$u = \sin(x)$$
$$\frac{du}{dx} = \cos(x)$$
$$du = \cos(x)dx$$

Now substitute:

$$\int \left(u^{-1/2} - u^{3/2} \right) du$$

And finish integrating as in Case 1.

Case 3: The powers of both sine and cosine are even and nonnegative

Here you convert the integrand into odd powers of cosines by using the following trig identities:

REMEMBER

$$\sin^2(x) = \frac{1 - \cos(2x)}{2} \quad \text{and} \quad \cos^2(x) = \frac{1 + \cos(2x)}{2}$$

Then you finish the problem as in Case 2. Here's an example:

$$\int \sin^4(x)\cos^2(x)dx$$
$$= \int \left(\sin^2(x) \right)^2 \cos^2(x)dx$$
$$= \int \left(\frac{1 - \cos(2x)}{2} \right)^2 \left(\frac{1 + \cos(2x)}{2} \right) dx$$
$$= \frac{1}{8} \int \left(1 - \cos(2x) - \cos^2(2x) + \cos^3(2x) \right) dx \quad \text{(It's just algebra!)}$$
$$= \frac{1}{8} \int 1\,dx - \frac{1}{8} \int \cos(2x)dx - \frac{1}{8} \int \cos^2(2x)dx + \frac{1}{8} \int \cos^3(2x)dx$$

The first in this string of integrals is a no-brainer; the second is a simple reverse rule with a little tweak for the 2; you do the third integral by using the $\cos^2(x)$ identity a second time; and the fourth integral is handled by following the steps in Case 2. Do it. Your final answer should be

$$\frac{1}{16}x - \frac{1}{64}\sin(4x) - \frac{1}{48}\sin^3(2x) + C$$

Secants and tangents

This section covers integrals — are you sitting down? — containing secants and tangents!

Case 1: The power of tangent is odd and positive

Integrate $\int \sqrt{\sec(x)} \tan^3(x)dx$.

1. **Lop off a secant-tangent factor and move it to the right.**

   First, rewrite the problem:
   $$\int \sqrt{\sec(x)}\tan^3(x)dx = \int \sec^{1/2}(x)\tan^3(x)dx.$$

   Taking a secant-tangent factor out of $\sec^{1/2}(x)\tan^3(x)$ may seem like squeezing blood from a turnip because $\sec^{1/2}(x)$ has a power less than $\sec^1(x)$, but it works:

   $$\int \sec^{1/2}(x)\tan^3(x)dx = \int \left(\sec^{-1/2}(x)\tan^2(x)\right)\sec(x)\tan(x)dx$$

2. **Convert the remaining tangents to secants with the tangent-secant version of the Pythagorean identity.**

   The Pythagorean identity is $\tan^2(x)+1=\sec^2(x)$, and thus $\tan^2(x)=\sec^2(x)-1$. Now make the switch.

   $$\int \left(\sec^{-1/2}(x)\tan^2(x)\right)\sec(x)\tan(x)dx$$
   $$= \int \left(\sec^{-1/2}(x)\left(\sec^2(x)-1\right)\right)\sec(x)\tan(x)dx$$
   $$= \int \left(\sec^{3/2}(x)-\sec^{-1/2}(x)\right)\sec(x)\tan(x)dx$$

3. **Solve by substitution with $u = \sec(x)$ and $du = \sec(x)\tan(x)dx$.**

   $$= \int \left(u^{3/2}-u^{-1/2}\right)du$$
   $$= \frac{2}{5}u^{5/2}-2u^{1/2}+C$$
   $$= \frac{2}{5}\sec^{5/2}(x)-2\sec^{1/2}(x)+C$$

Case 2: The power of secant is even and positive

Find $\int \sec^4(x)\tan^4(x)dx$.

1. **Lop off a $\sec^2(x)$ factor and move it to the right.**
   $$= \int \sec^2(x)\tan^4(x)\sec^2(x)dx$$

2. **Convert the remaining secants to tangents with the Pythagorean identity, $\sec^2(x)=\tan^2(x)+1$.**
   $$= \int \left(\tan^2(x)+1\right)\tan^4(x)\sec^2(x)dx$$
   $$= \int \left(\tan^6(x)+\tan^4(x)\right)\sec^2(x)dx$$

3. **Solve by substitution, where $u = \tan(x)$ and $du = \sec^2(x)dx$.**

$$= \int \left(u^6 + u^4 \right) du$$

$$= \frac{1}{7} u^7 + \frac{1}{5} u^5 + C$$

$$= \frac{1}{7} \tan^7(x) + \frac{1}{5} \tan^5(x) + C$$

Case 3: The power of tangent is even and positive and there are no secant factors

Integrate $\int \tan^6(x)dx$.

1. **Convert one $\tan^2(x)$ factor to secants using the Pythagorean identity, $\tan^2(x) = \sec^2(x) - 1$.**

$$= \int \tan^4(x) \left(\sec^2(x) - 1 \right) dx$$

2. **Distribute and split up the integral.**

$$= \int \tan^4(x) \sec^2(x)dx - \int \tan^4(x)dx$$

3. **Solve the first integral like in Step 3 of Case 2 for secants and tangents.**

   You should get $\int \tan^4(x) \sec^2(x)dx = \frac{1}{5} \tan^5(x) + C$.

4. **For the second integral from Step 2, go back to Step 1 and repeat the process.**

   For this piece of the problem, you get

$$-\int \tan^4(x)dx = -\int \tan^2(x) \sec^2(x)dx + \int \tan^2(x)dx.$$

5. **Repeat Step 3 for $-\int \tan^2(x) \sec^2(x)dx$ (using Case 2, Step 3) for secants and tangents again).**

$$-\int \tan^2(x) \sec^2(x)dx = -\frac{1}{3} \tan^3(x) + C$$

6. **Use the Pythagorean identity to convert the $\int \tan^2(x)dx$ from Step 4 into $\int \sec^2(x)dx - \int 1 dx$.**

   Both of these integrals can be done with simple reverse differentiation rules. After collecting all these pieces — piece 1 from Step 3, piece 2 from Step 5, and pieces 3 and 4 from Step 6 — your final answer should be $\int \tan^6(x)dx = \frac{1}{5} \tan^5(x) - \frac{1}{3} \tan^3(x) + \tan(x) - x + C$.

Cosecants and cotangents

Cosecant and cotangent integrals work exactly like the three cases for secants and tangents — you just use a different form of the Pythagorean identity: $1 + \cot^2(x) = \csc^2(x)$. Try this one:

Integrate $\int \dfrac{\cot^3(x)}{\sqrt{\csc(x)}}\, dx$. If you get $-2\sin^{1/2}(x) - \frac{2}{3}\csc^{3/2}(x) + C$, pass "Go" and collect \$200.

If you get a secant-tangent or a cosecant-cotangent problem that doesn't fit any of the cases discussed in the previous section or if you're otherwise stumped by a problem, try converting it to sines and cosines and solving it with one of the sine-cosine methods or with identities like

$$\sin^2(x) + \cos^2(x) = 1 \text{ and } \cos^2(x) = \frac{1 + \cos(2x)}{2}.$$

Trigonometric Substitution

With the trigonometric substitution method, you can do integrals containing radicals of the following forms: $\sqrt{u^2 + a^2}$, $\sqrt{a^2 - u^2}$, and $\sqrt{u^2 - a^2}$ (as well as powers of those roots), where a is a constant and u is an expression containing x. For instance, $\sqrt{3^2 - x^2}$ is of the form $\sqrt{a^2 - u^2}$.

I've got some silly mnemonic tricks to help you keep the three cases of this method straight. (Remember, with mnemonic devices, silly works.) First, the three cases involve three trig functions, *tangent, sine,* and *secant.* Their initial letters, *t, s,* and *s,* are the same letters as the initial letters of the name of this technique, *trigonometric substitution.* Pretty nice, eh?

Table 10-1 shows how these three trig functions pair up with the radical forms listed in the opening paragraph.

Table 10-1	A Totally Radical Table
$\tan(\theta) \longleftrightarrow \sqrt{u^2 + a^2}$	
$\sin(\theta) \longleftrightarrow \sqrt{a^2 - u^2}$	
$\sec(\theta) \longleftrightarrow \sqrt{u^2 - a^2}$	

To keep these pairings straight, note that the plus sign in $\sqrt{u^2+a^2}$ looks like a little t for *tangent,* and that the other two forms, $\sqrt{a^2-u^2}$ and $\sqrt{u^2-a^2}$, contain a *subtraction* sign — s is for *sine* and *secant.* To memorize what sine and secant pair up with, note that $\sqrt{a^2-u^2}$ begins with the letter a, and it's a *sin* to call someone an *ass*. Okay, I admit this is pretty weak. If you can come up with a better mnemonic, use it!

Case 1: Tangents

Find $\int \dfrac{dx}{\sqrt{9x^2+4}}$. First, note that this can be rewritten as

$\int \dfrac{dx}{\sqrt{(3x)^2+2^2}}$, so it fits the form $\sqrt{u^2+a^2}$, where $u=3x$ and $a=2$.

1. **Draw a right triangle — basically a *SohCahToa* triangle — where $\tan(\theta)$ equals $\dfrac{u}{a}$, which is $\dfrac{3x}{2}$.**

   Because you know that $\tan(\theta)=\dfrac{O}{A}$ (from *SohCahToa*), your triangle should have $3x$ as O, the side *opposite* the angle θ, and 2 as A, the *adjacent* side. Then, by the Pythagorean Theorem, the length of the hypotenuse automatically equals your radical, $\sqrt{(3x)^2+2^2}$, or $\sqrt{9x^2+4}$. See Figure 10-5.

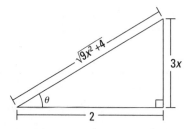

Figure 10-5: A *SohCahToa* triangle for the $\sqrt{u^2+a^2}$ case.

2. **Solve $\tan(\theta)=\dfrac{3x}{2}$ for x, differentiate, and solve for dx.**

$$\frac{3x}{2}=\tan(\theta)$$
$$3x=2\tan(\theta)$$
$$x=\frac{2}{3}\tan(\theta)$$

$$\frac{dx}{d\theta} = \frac{2}{3}\sec^2(\theta)$$

$$dx = \frac{2}{3}\sec^2(\theta)d\theta$$

3. **Find which trig function is represented by the radical over the a and then solve for the radical.**

   Look at the triangle in Figure 10-5. The radical is the hypotenuse and a is 2, the *adjacent* side, so $\dfrac{\sqrt{9x^2+4}}{2}$ is $\dfrac{H}{A}$, which equals *secant*. So $\sec(\theta) = \dfrac{\sqrt{9x^2+4}}{2}$, and thus $\sqrt{9x^2+4} = 2\sec(\theta)$.

4. **Use the results from Steps 2 and 3 to make substitutions in the original problem, then integrate.**

   From Steps 2 and 3 you have $dx = \dfrac{2}{3}\sec^2(\theta)d\theta$ and $\sqrt{9x^2+4} = 2\sec(\theta)$. Now you can finally do the integration.

$$\int \frac{dx}{\sqrt{9x^2+4}} = \int \frac{\frac{2}{3}\sec^2(\theta)d\theta}{2\sec(\theta)}$$

$$= \frac{1}{3}\int \sec(\theta)d\theta$$

$$= \frac{1}{3}\ln\left|\sec(\theta)+\tan(\theta)\right| + C \quad \text{(something you should memorize)}$$

5. **Substitute the x expressions from Steps 1 and 3 back in for $\sec(\theta)$ and $\tan(\theta)$. You can also get the expressions from the triangle in Figure 10-5.**

$$= \frac{1}{3}\ln\left|\frac{\sqrt{9x^2+4}}{2} + \frac{3x}{2}\right| + C$$

$$= \frac{1}{3}\ln\left|\frac{\sqrt{9x^2+4}+3x}{2}\right| + C$$

$$= \frac{1}{3}\ln\left|\sqrt{9x^2+4}+3x\right| - \frac{1}{3}\ln 2 + C \quad \left(\begin{array}{l}\text{by the log of a quotient}\\ \text{rule and distributing the } \frac{1}{3}\end{array}\right)$$

$$= \frac{1}{3}\ln\left|\frac{\sqrt{9x^2+4}}{2} + \frac{3x}{2}\right| + C \quad \left(\text{because } \frac{1}{3}\ln 2 + C \text{ is just a constant}\right)$$

For all three cases in trigonometric substitution, Step 1 always involves drawing a triangle in which the trig function in question equals $\frac{u}{a}$:

$$\text{Case 1 is } \tan(\theta) = \frac{u}{a}$$

$$\text{Case 2 is } \sin(\theta) = \frac{u}{a}$$

$$\text{Case 3 is } \sec(\theta) = \frac{u}{a}$$

For all three cases, Step 3 always involves putting the radical over the *a*. The three cases are given below, but you don't need to memorize the trig functions in this list because you'll know which one you've got by just looking at the triangle—assuming you know *SohCahToa* and the reciprocal trig functions. I've left out what goes under the radicals because by the time you're doing Step 3, you've already got the right radical expression.

$$\text{Case 1 is } \sec(\theta) = \frac{\sqrt{}}{a}$$

$$\text{Case 2 is } \cos(\theta) = \frac{\sqrt{}}{a}$$

$$\text{Case 3 is } \tan(\theta) = \frac{\sqrt{}}{a}$$

Just remember $\frac{u}{a}$ for Step 1 and $\frac{\sqrt{}}{a}$ for Step 3.

Case 2: Sines

Integrate $\int \frac{dx}{x^2\sqrt{16-x^2}}$, rewriting it first as $\int \frac{dx}{x^2\sqrt{4^2-x^2}}$ so that it fits the form $\sqrt{a^2-u^2}$, where $a = 4$ and $u = x$.

1. **Draw a right triangle where** $\sin(\theta) = \frac{u}{a}$, **which is** $\frac{x}{4}$.

   Sine equals $\frac{O}{H}$, so the *opposite* side is *x* and the *hypotenuse* is 4. The length of the adjacent side is then automatically equal to your radical, $\sqrt{16-x^2}$. See Figure 10-6.

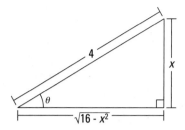

Figure 10-6: A *SohCahToa* triangle for the $\sqrt{a^2 - u^2}$ case.

2. **Solve $\sin(\theta) = \frac{x}{4}$ for x, differentiate, and solve for dx.**

$$\frac{x}{4} = \sin(\theta)$$

$$x = 4\sin(\theta)$$

$$\frac{dx}{d\theta} = 4\cos(\theta)$$

$$dx = 4\cos(\theta)d\theta$$

3. **Find which trig function equals the radical over the a, and then solve for the radical.**

Look at the triangle in Figure 10-6. The radical, $\sqrt{16 - x^2}$, over the a, 4, is $\frac{A}{H}$, which you know from *SohCahToa* equals *cosine*. So that gives you

$$\cos(\theta) = \frac{\sqrt{16 - x^2}}{4}, \text{ and thus}$$

$$\sqrt{16 - x^2} = 4\cos(\theta)$$

4. **Use the results from Steps 2 and 3 to make substitutions in the original problem, then integrate.**

Note that in this particular problem, you have to make three substitutions, not just two like in the first example. From Steps 2 and 3 you've got $x = 4\sin(\theta)$, $dx = 4\cos(\theta)d\theta$, and $\sqrt{16 - x^2} = 4\cos(\theta)$, so

$$\int \frac{dx}{x^2\sqrt{16 - x^2}} = \int \frac{4\cos(\theta)d\theta}{\left(4\sin(\theta)\right)^2 4\cos(\theta)}$$

$$= \int \frac{d\theta}{16\sin^2(\theta)}$$

$$= \frac{1}{16} \int \csc^2(\theta) d\theta$$

$$= -\frac{1}{16} \cot(\theta) + C$$

5. **The triangle shows that** $\cot(\theta) = \frac{\sqrt{16-x^2}}{x}$. **Substitute back for your final answer.**

$$= -\frac{1}{16} \cdot \frac{\sqrt{16-x^2}}{x} + C$$

$$= -\frac{\sqrt{16-x^2}}{16x} + C$$

Case 3: Secants

In the interest of space — and sanity — let's skip this case. But you won't have any trouble with it, because all the steps are basically the same as in Cases 1 and 2. Try this one.

Integrate $\int \frac{\sqrt{x^2-9}}{x} \, dx$. I'll get you started. In Step 1, you draw a triangle, where $\sec(\theta) = \frac{u}{a}$, that's $\frac{x}{3}$. Now take it from there.

You should get: $\sqrt{x^2-9} - 3\arctan\left(\frac{\sqrt{x^2-9}}{3}\right) + C$.

Partial Fractions

You use the partial fractions method to integrate rational functions like $\frac{6x^2+3x-2}{x^3+2x^2}$. The basic idea involves "unadding" a fraction. Adding works like this: $\frac{1}{2} + \frac{1}{3} = \frac{5}{6}$. So you can "unadd" $\frac{5}{6}$ by splitting it up into $\frac{1}{2}$ plus $\frac{1}{3}$. This is what you do with the partial fraction technique except that you do it with complicated rational functions instead of ordinary fractions.

Case 1: The denominator contains only linear factors

Integrate $\int \frac{5}{x^2+x-6} \, dx$. This is Case 1 because the factored denominator (see Step 1) contains only *linear* factors.

1. **Factor the denominator.**

$$\frac{5}{x^2 + x - 6} = \frac{5}{(x-2)(x+3)}$$

2. **Break up the fraction on the right into a sum of fractions, where each factor of the denominator in Step 1 becomes the denominator of a separate fraction. Then put unknowns in the numerator of each fraction.**

$$\frac{5}{(x-2)(x+3)} = \frac{A}{(x-2)} + \frac{B}{(x+3)}$$

3. **Multiply both sides of this equation by the denominator of the left side.**

$$5 = A(x+3) + B(x-2)$$

4. **Take the roots of the linear factors and plug them — one at a time — into x in the equation from Step 3, and solve for the unknowns.**

If $x = 2$,   If $x = -3$,

$5 = A(2+3) + B(2-2)$ $5 = A(-3+3) + B(-3-2)$

$5 = 5A$ $5 = -5B$

$A = 1$ $B = -1$

5. **Plug these results into the A and B in the equation from Step 2.**

$$\frac{5}{(x-2)(x+3)} = \frac{1}{(x-2)} + \frac{-1}{(x+3)}$$

6. **Split up the original integral into the partial fractions from Step 5 and you're home free.**

$$\int \frac{5}{x^2 + x - 6}\,dx = \int \frac{1}{x-2}\,dx + \int \frac{-1}{x+3}\,dx$$

$$= \ln|x-2| - \ln|x+3| + C$$

$$= \ln\left|\frac{x-2}{x+3}\right| + C \quad \text{(the log of a quotient rule)}$$

Case 2: The denominator contains unfactorable quadratic factors

Sometimes you can't factor a denominator all the way down to linear factors because some quadratics can't be factored.

Here's a problem: Integrate $\int \dfrac{5x^3+9x-4}{x(x-1)(x^2+4)}\, dx$.

1. **Factor the denominator.**

   It's already done! Note that x^2+4 can't be factored.

2. **Break up the fraction into a sum of "partial fractions."**

   If you have an irreducible quadratic factor (like the x^2+4), the numerator for that partial fraction needs two unknowns in the form $Px + Q$.

   $$\dfrac{5x^3+9x-4}{x(x-1)(x^2+4)} = \dfrac{A}{x} + \dfrac{B}{x-1} + \dfrac{Cx+D}{x^2+4}$$

3. **Multiply both sides of this equation by the left-side denominator.**

$$5x^3+9x-4 = A(x-1)(x^2+4)+B(x)(x^2+4)+(Cx+D)x(x-1)$$

4. **Take the roots of the linear factors and plug them — one at a time — into x in the equation from Step 3, and then solve.**

If $x=0$,	If $x=1$,
$-4 = -4A$	$10 = 5B$
$A = 1$	$B = 2$

   Unlike in the Case 1 example, you can't solve for all the unknowns by plugging in the roots of the linear factors, so you have more work to do.

5. **Plug into the Step 3 equation the known values of A and B and any two values for x not used in Step 4 to get a system of two equations in C and D.**

   $A = 1$ and $B = 2$, so

If $x = -1$,	If $x = 2$,
$-18 = -10 - 10 - 2C + 2D$	$54 = 8 + 32 + 4C + 2D$
$2 = -2C + 2D$	$14 = 4C + 2D$
$1 = -C + D$	$7 = 2C + D$

6. **Solve the system:** $1 = -C + D$ **and** $7 = 2C + D$.

   You should get $C = 2$ and $D = 3$.

7. **Split up the original integral and integrate.**

Using the values obtained in Steps 4 and 6, $A = 1$, $B = 2$, $C = 2$, and $D = 3$, and the equation from Step 2, you can split up the original integral into three pieces:

$$\int \frac{5x^3 + 9x - 4}{x(x-1)(x^2+4)} \, dx = \int \frac{1}{x} \, dx + \int \frac{2}{x-1} \, dx + \int \frac{2x+3}{x^2+4} \, dx$$

And with basic algebra, you can split up the third integral on the right into two pieces, resulting in the final partial fraction decomposition:

$$\int \frac{5x^3 + 9x - 4}{x(x-1)(x^2+4)} \, dx = \int \frac{1}{x} \, dx + \int \frac{2}{x-1} \, dx + \int \frac{2x}{x^2+4} \, dx + \int \frac{3}{x^2+4} \, dx$$

The first two integrals are easy. For the third, you use substitution with $u = x^2 + 4$ and $du = 2x \, dx$. The fourth is done with the arctangent rule which you should memorize: $\int \frac{dx}{a^2+x^2} = \frac{1}{a} \arctan\left(\frac{x}{a}\right) + C$.

$$\int \frac{5x^3 + 9x - 4}{x(x-1)(x^2+4)} \, dx = \ln|x| + 2\ln|x-1| + \ln|x^2+4| + \frac{3}{2}\arctan\left(\frac{x}{2}\right) + C$$

$$= \ln\left|x(x-1)^2(x^2+4)\right| + \frac{3}{2}\arctan\left(\frac{x}{2}\right) + C$$

Case 3: The denominator contains repeated factors

If the denominator contains any repeated factors (linear or quadratic), like $(x+5)^4$, here's what you do: Say you want to

integrate $\int \frac{1}{x^2(x-1)^3} \, dx$. The x in the denominator has a

power of 2, so you get 2 partial fractions for the x (for the powers of 1 and 2); the $(x-1)$ has a power of 3, so you get 3 partial fractions for that factor (for the powers 1, 2, and 3). Here's the general form for the partial fraction decomposition:

$\frac{A}{x} + \frac{B}{x^2} + \frac{C}{(x-1)} + \frac{D}{(x-1)^2} + \frac{E}{(x-1)^3}$. Here's another one. You break up $\dfrac{4x^2 - x^2 + 8}{(2x-3)^2(x^2+1)^2}$ into $\dfrac{A}{(2x-3)} + \dfrac{B}{(2x-3)^2} +$

$\dfrac{Cx+D}{(x^2+1)} + \dfrac{Ex+F}{(x^2+1)^2}$. I'm skipping the solutions for these examples.

The method's the same as in Cases 1 and 2 above — just messier.

Equating coefficients

Here's another method for finding your capital letter unknowns that you should have in your bag of tricks. Say you get the following for your Step 3 equation:

$$2x^3 + x^2 - 5x + 4 = (Ax+B)(x^2+1) + (Cx+D)(x^2+2x+2)$$

This equation has no linear factors, so you can't plug in the roots to get the unknowns. Instead, expand the right side of the equation:

$$2x^3 + x^2 - 5x + 4 = Ax^3 + Ax + Bx^2 + B + Cx^3 + 2Cx^2 + 2Cx + Dx^2 + 2Dx + 2D$$

And collect like terms:

$$2x^3 + x^2 - 5x + 4 = (A+C)x^3 + (B+2C+D)x^2 + (A+2C+2D)x + (B+2D)$$

Then equate the coefficients of like terms from the left and right sides of the equation:

$$2 = A + C$$
$$1 = B + 2C + D$$
$$-5 = A + 2C + 2D$$
$$4 = B + 2D$$

You then solve this system of simultaneous equations to get A, B, C, and D.

In a nutshell, you have three ways to find your A, B, C, etc.: 1) Plugging in the roots of the linear factors of the denominator if there are any, 2) Plugging in other values of x and solving the resulting system of equations, and 3) Equating the coefficients of like terms. With practice, you'll get good at combining these methods to find your unknowns quickly.

Chapter 11

Using the Integral to Solve Problems

● ●

In This Chapter

▶ Adding up the area between curves

▶ Figuring out volumes of weird shapes

▶ Finding arc length

● ●

A s I say in Chapter 8, integration is basically adding up small pieces of something to get the total for the whole thing — *really* small pieces, actually, *infinitely* small pieces. Thus, the integral

$$\int_{5\,\text{sec.}}^{20\,\text{sec.}} little\ piece\ of\ distance$$

tells you to add up all the little pieces of distance traveled during the 15-second interval from 5 to 20 seconds to get the total distance traveled during that interval. In all problems, the little piece after the integration symbol is always an expression in x (or other variable). For the above integral, for instance, the little piece of distance might be given by, say, x^2dx. Then the definite integral

$$\int_{5}^{20} x^2\,dx$$

would give you the total distance traveled. Your main challenge in this chapter is simply to come up with the algebraic expression for the little pieces you're adding up. Before we begin the adding-up problems, I want to cover a couple other integration topics.

The Mean Value Theorem for Integrals and Average Value

The best way to understand the Mean Value Theorem for Integrals is with a diagram — look at Figure 11-1.

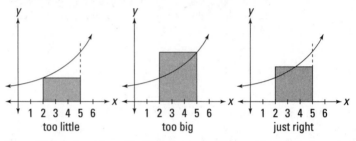

Figure 11-1: A visual "proof" of the mean value theorem for integrals.

The graph on the left in Figure 11-1 shows a rectangle whose area is clearly *less than* the area under the curve between 2 and 5. This rectangle has a height equal to the lowest point on the curve in the interval from 2 to 5. The middle graph shows a rectangle whose height equals the highest point on the curve. Its area is clearly *greater than* the area under the curve. By now you're thinking, "Isn't there a rectangle taller than the short one and shorter than the tall one whose area is *the same as* the area under the curve?" Of course. And this rectangle obviously crosses the curve somewhere in the interval. This so-called "mean value rectangle," shown on the right, basically sums up the Mean Value Theorem for Integrals. It's really just common sense. But here's the mumbo jumbo.

The Mean Value Theorem for Integrals: If $f(x)$ is a continuous function on the closed interval $[a, b]$, then there exists a number c in the closed interval such that

$$\int_a^b f(x)\,dx = f(c)\cdot(b-a)$$

The theorem basically just guarantees the existence of the mean value rectangle.

The area of the mean value rectangle — which is the same as the area under the curve — equals *length* times *width*, or *base* times *height,* right? So, if you divide its area, $\int_a^b f(x)dx$, by its base, $(b-a)$, you get its height, $f(c)$. This height is the *average value* of the function over the interval in question.

Average Value: The *average value* of a function $f(x)$ over a closed interval $[a, b]$ is

$$\frac{1}{b-a}\int_a^b f(x)dx$$

which is the height of the mean value rectangle.

Here's an example: What's the average speed of a car between $t = 9$ seconds and $t = 16$ seconds whose speed in *feet per second* is given by the function $f(t) = 30\sqrt{t}$? According to the definition of average value, this average speed is given by

$$\frac{1}{16-9}\int_9^{16} 30\sqrt{t}\ dt.$$

1. **Determine the area under the curve between 9 and 16.**

$$\int_9^{16} 30\sqrt{t}\ dt$$

$$= 30\left[\frac{2}{3}t^{3/2}\right]_9^{16}$$

$$= 30\left(\frac{128}{3} - \frac{54}{3}\right)$$

$$= 740$$

This area, by the way, is the total distance traveled from 9 to 16 seconds. Do you see why? Consider the mean value rectangle to answer this question. Its height is a speed (because the function values, or heights, are speeds) and its base is an amount of time, so its area is *speed* times *time* which equals *distance*.

2. **Divide this area, total distance, by the time interval from 9 to 16, namely 7.**

$$Average\ speed = \frac{total\ distance}{total\ time} = \frac{740\ feet}{7\ seconds}$$

$$\approx 105.7\ feet\ per\ second$$

The definition of average value tells you to multiply the total area by $\frac{1}{b-a}$, which in this problem is $\frac{1}{16-9}$, or $\frac{1}{7}$. But because dividing by 7 is the same as multiplying by $\frac{1}{7}$, you can divide like I do in this step. It makes more sense to think about these problems in terms of division: *area* equals *base* times *height*, so the height of the mean value rectangle equals its area *divided* by its base.

The Area between Two Curves

This is the first of seven topics in this chapter where your task is to come up with an expression for a little bit of something and then add up the bits by integrating. For this first problem type, the little bit is a narrow rectangle that sits on one curve and goes up to another. Here's an example: Find the area between $y = 2 - x^2$ and $y = \frac{1}{2}x$ from $x = 0$ to $x = 1$. See Figure 11-2.

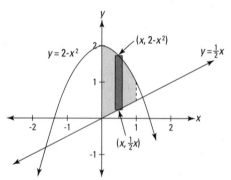

Figure 11-2: The area between $y = 2 - x^2$ and $y = \frac{1}{2}x$ from $x = 0$ to $x = 1$.

To get the height of the representative rectangle in Figure 11-2, subtract the y-coordinate of its bottom from the y-coordinate of its top — that's $(2 - x^2) - \frac{1}{2}x$. Its base is the infinitesimal *dx*. So, because *area* equals *height* times *base*,

$$\text{Area of representative rectangle} = \left((2 - x^2) - \frac{1}{2}x \right) dx$$

Now add up the areas of all rectangles from 0 to 1 by integrating:

$$\int_0^1 \left((2-x^2) - \frac{1}{2}x \right) dx$$

$$= \left[2x - \frac{1}{3}x^3 - \frac{1}{4}x^2 \right]_0^1 \text{ (power rule for all 3 pieces)}$$

$$= \left(2 - \frac{1}{3} - \frac{1}{4} \right) - (0 - 0 - 0)$$

$$= \frac{17}{12} \text{ square units}$$

To make things a little more twisted, in the next problem the curves cross (see Figure 11-3). When this happens, you have to split the total shaded area into two separate regions before integrating. Find the area between $\sqrt[3]{x}$ and x^3 from $x = 0$ to $x = 2$.

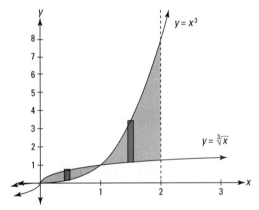

Figure 11-3: Who's on top?

1. **Determine where the curves cross.**

   They cross at (1, 1), so you've got two separate regions — one from 0 to 1 and another from 1 to 2.

2. **Figure the area of the region on the left.**

   For this region, $\sqrt[3]{x}$ is above x^3. So the height of a representative rectangle is $\sqrt[3]{x} - x^3$, its area is *height* times *base,* or $\left(\sqrt[3]{x} - x^3 \right) dx$, and the area of the region is, therefore,

$$\int_0^1 \left(\sqrt[3]{x} - x^3\right) dx$$

$$= \left[\frac{3}{4}x^{4/3} - \frac{1}{4}x^4\right]_0^1$$

$$= \left(\frac{3}{4} - \frac{1}{4}\right) - (0 - 0)$$

$$= \frac{1}{2}$$

3. **Figure the area of the region on the right.**

   Now, x^3 is above $\sqrt[3]{x}$, so the height of a rectangle is $x^3 - \sqrt[3]{x}$ and thus you've got

$$\int_1^2 \left(x^3 - \sqrt[3]{x}\right) dx$$

$$= \left[\frac{1}{4}x^4 - \frac{3}{4}x^{4/3}\right]_1^2$$

$$= \left(4 - \frac{3}{2}\sqrt[3]{2}\right) - \left(\frac{1}{4} - \frac{3}{4}\right)$$

$$= 4.5 - 1.5\sqrt[3]{2}$$

4. **Add up the areas of the two regions to get total area.**

$$0.5 + 4.5 - 1.5\sqrt[3]{2} \approx 3.11 \text{ square units}$$

Volumes of Weird Solids

Integration enables you to calculate the volumes of shapes that you can't get with basic geometry formulas.

The meat-slicer method

This metaphor is actually quite accurate. Picture a hunk of meat being cut into very thin slices on one of those deli meat slicers. That's the basic idea here. You slice up a three-dimensional shape, then add up the volumes of the slices to determine the total volume.

Here's a problem: What's the volume of the solid whose length runs along the x-axis from 0 to π and whose cross sections perpendicular to the x-axis are equilateral triangles such that the midpoints of their bases lie on the x-axis and their top vertices are on the curve $y = \sin(x)$? Is that a mouthful or what? This problem is almost harder to describe and to picture than it is to do. Take a look at this thing in Figure 11-4.

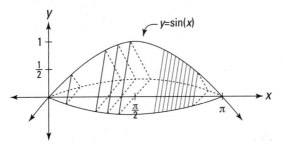

Figure 11-4: One weird hunk of pastrami.

So what's the volume?

1. **Determine the area of any old cross section.**

   Each cross section is an equilateral triangle with a height of $\sin(x)$. If you do the geometry, you'll see that its base is $\frac{2\sqrt{3}}{3}$ times its height, or $\frac{2\sqrt{3}}{3} \cdot \sin(x)$ (Hint: Half of an equilateral triangle is a 30°-60°-90° triangle). So, the triangle's area, given by $A = \frac{1}{2}(b)(h)$ is $\frac{1}{2}\left(\frac{2\sqrt{3}}{3} \cdot \sin(x)\right)\sin(x)$, or $\frac{\sqrt{3}}{3}\sin^2 x$.

2. **Find the volume of a representative slice.**

   The volume of a slice is just its cross-sectional area times its infinitesimal thickness, dx. So you've got its volume:

   $$\text{Volume of representative slice} = \frac{\sqrt{3}}{3}\sin^2(x)\,dx$$

3. Add up the volumes of the slices from 0 to π by integrating.

$$\int_0^\pi \frac{\sqrt{3}}{3} \sin^2(x)\,dx$$

$$= \frac{\sqrt{3}}{3} \int_0^\pi \sin^2(x)\,dx$$

$$= \frac{\sqrt{3}}{3} \int_0^\pi \frac{1-\cos(2x)}{2}\,dx \quad \left(\begin{array}{l} \text{trig integrals with sines} \\ \text{and cosines, Case 3,} \\ \text{from Chapter 10} \end{array} \right)$$

$$= \frac{\sqrt{3}}{6} \left(\int_0^\pi 1\,dx - \int_0^\pi \cos(2x)\,dx \right)$$

$$= \frac{\sqrt{3}}{6} \left(\left[x \right]_0^\pi - \left[\frac{\sin(2x)}{2} \right]_0^\pi \right)$$

$$= \frac{\sqrt{3}}{6} \left(\pi - 0 - \left(\frac{\sin(2\pi)}{2} - \frac{\sin(0)}{2} \right) \right)$$

$$= \frac{\sqrt{3}}{6} \left(\pi - 0 - (0-0) \right)$$

$$= \frac{\pi\sqrt{3}}{6}$$

$$\approx 0.91 \, cubic \, units$$

The disk method

This technique is a special case of the meat slicer method that you use when the cross-sections are all circles. Find the volume of the solid — between $x = 2$ and $x = 3$ — generated by rotating the curve $y = e^x$ about the x-axis. See Figure 11-5.

1. Determine the area of any old cross section or representative disk.

Each cross section is a circle with a radius of e^x. So, its area is given by the formula for the area of a circle, $A = \pi r^2$. Plugging e^x into r gives you

$$A = \pi \left(e^x \right)^2 = \pi e^{2x}$$

2. **Tack on a *dx* to get the volume of an infinitely thin representative disk.**

$$\overbrace{\text{Volume of disk} = \pi e^{2x}}^{area} \cdot \overbrace{dx}^{thickness}$$

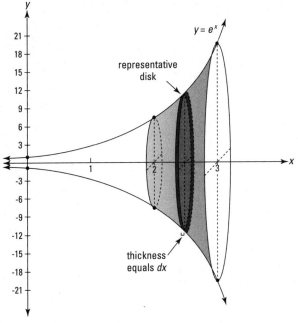

Figure 11-5: A sideways stack of disks.

3. **Add up the volumes of the disks from 2 to 3 by integrating.**

$$\text{Total volume} = \int_{2}^{3} \pi e^{2x} dx$$

$$= \pi \int_{2}^{3} e^{2x} dx$$

$$= \frac{\pi}{2} \left[e^{2x} \right]_{2}^{3} \left(\text{by substitution with } u = 2x \text{ and } du = 2dx \right)$$

$$= \frac{\pi}{2} \left(e^{6} - e^{4} \right)$$

$$\approx 548 \text{ cubic units}$$

The washer method

The only difference between the washer method and the disk method is that now each slice has a hole in its middle that you have to subtract. Take the area bounded by $y = x^2$ and $y = \sqrt{x}$, and generate a solid by revolving that area about the x-axis. See Figure 11-6.

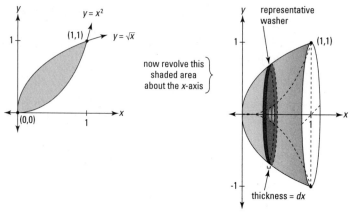

Figure 11-6: A sideways stack of washers — just add up the volumes of all the washers.

So what's the volume of this bowl-like shape?

1. **Determine where the two curves intersect.**

   It should take very little trial and error to see that $y = x^2$ and $y = \sqrt{x}$ intersect at $x = 0$ and $x = 1$. So the solid in question spans the interval on the x-axis from 0 to 1.

2. **Figure the area of a thin cross-sectional washer.**

   Each slice has the shape of a washer — see Figure 11-7 — so its area equals the area of the entire circle minus the area of the hole.

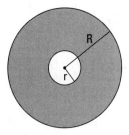

Figure 11-7: The shaded area equals $\pi R^2 - \pi r^2$: The *whole* minus the *hole* — get it?

The area of the circle minus the hole is $\pi R^2 - \pi r^2$, where R is the outer radius (the big radius) and r is the radius of the hole (the little radius). For this problem, the outer radius is $\sqrt{x}$ and the radius of the hole is x^2, so that gives you

$$A = \pi\left(\sqrt{x}\right)^2 - \pi\left(x^2\right)^2$$
$$= \pi x - \pi x^4$$

3. **Multiply this area by the thickness, *dx*, to get the volume of a representative washer.**

$$Volume = \left(\pi x - \pi x^4\right)dx$$

4. **Add up the volumes of the even-thinner-than-paper-thin washers from 0 to 1 by integrating.**

$$Total\ volume = \int_0^1 \left(\pi x - \pi x^4\right)dx$$
$$= \pi\int_0^1 \left(x - x^4\right)dx$$
$$= \pi\left[\frac{1}{2}x^2 - \frac{1}{5}x^5\right]_0^1$$
$$= \pi\left[\left(\frac{1}{2} - \frac{1}{5}\right) - (0-0)\right]$$
$$= \frac{3}{10}\pi$$
$$\approx 0.94\ cubic\ units$$

The matryoshka doll method

Now you're going to cut up a solid into thin concentric cylinders and then add up the volumes of all the cylinders. It's kinda like how those nested Russian (matryoshka) dolls fit inside each other. Or imagine a soup can that somehow has many paper labels, each one covering the one beneath it. Or picture one of those clothes de-linters with the sticky papers you peel off. Each soup can label or piece of sticky paper is a cylindrical shell — before you tear it off, of course. After you tear it off, it's an ordinary rectangle.

Here's a problem: A solid is generated by taking the area bounded by the x-axis, the lines $x = 2$, $x = 3$, and $y = e^x$, and then revolving it about the y-axis. What's its volume? See Figure 11-8.

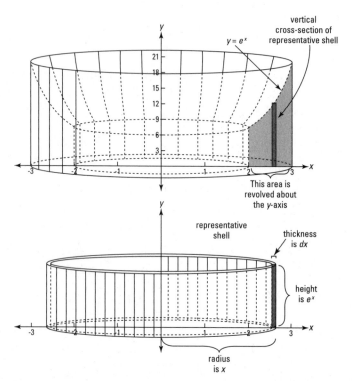

Figure 11-8: A shape sort of like the Roman Coliseum and one of its representative shells.

1. **Determine the area of a representative cylindrical shell.**

   When picturing a representative shell, focus on one that's in no place in particular. Figure 11-8 shows such a generic shell. Its radius is unknown, x, and its height is the height of the curve at x, namely e^x. If, instead, you use a special shell like the outermost shell with a radius of 3, you're more likely to make the mistake of thinking that a representative shell has some *known* radius like 3 or a *known* height like e^3. Both the radius and the height are *unknown*. (This advice also applies to disk and washer problems.)

   Each representative shell, like the soup can label or the sticky sheet from a de-linter, is just a rectangle whose area is, of course, *length* times *width*. The rectangular soup can label goes all the way around the can, so its length is the circumference of the can, namely $2\pi r$; the width of the label is the height of the can. So now you've got the general formula for the area of a representative shell:

   $$Area = length \cdot width$$
   $$= 2\pi r \cdot h$$

   For the current problem, you plug in x for the radius and e^x for the height, giving you the area of a representative shell:

   $$Area\ of\ shell = 2\pi x \cdot e^x$$

2. **Multiply the area by the thickness of the shell, *dx*, to get its volume.**

   $$Volume\ of\ representative\ shell = 2\pi x e^x dx$$

3. **Add up the volumes of all the shells from 2 to 3 by integrating.**

   $$Total\ volume = \int_2^3 2\pi x e^x\, dx$$
   $$= 2\pi \int_2^3 x e^x\, dx$$
   $$= 2\pi \left[x e^x - e^x \right]_2^3 \text{ (integration by parts)}$$

$$= 2\pi\left(3e^3 - e^3 - \left(2e^2 - e^2\right)\right)$$
$$= 2\pi\left(2e^3 - e^2\right)$$
$$\approx 206 \; cubic \; units$$

With the meat-slicer, disk, and washer methods, it's usually pretty obvious what the limits of integration should be (recall that the *limits of integration* are, for example, the 1 and 5 in $\int_1^5$).

With cylindrical shells, however, it's not always as clear. A tip: You integrate from the *right* edge of the smallest cylinder to the *right* edge of the biggest cylinder (like from 2 to 3 in the previous problem). And note that you never integrate from the left edge to the right edge of the biggest cylinder (like from –3 to 3).

Arc Length

So far in this chapter, you've added up the areas of thin rectangles to get total area, and the volumes of thin slices or thin cylinders to get total volume. Now, you're going to add up minute lengths along a curve, an "arc," to get the whole length. The idea is to divide a length of curve into small sections, figure the length of each, and then add them all up. Figure 11-9 shows how each section of a curve is approximated by the hypotenuse of a tiny right triangle.

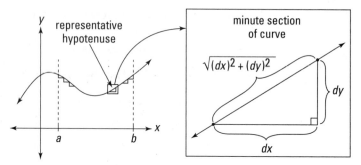

Figure 11-9: The Pythagorean Theorem, $a^2 + b^2 = c^2$, is the key to the arc length formula.

You can imagine that as you zoom in further and further, dividing the curve into more and more sections, the minute sections get straighter and straighter and the hypotenuses become better and better approximations of the curve. So, all you have to do is add up all the hypotenuses along the curve between your start and finish points. The lengths of the legs of each infinitesimal triangle are dx and dy, and thus the length of the hypotenuse — given by the Pythagorean Theorem — is

$$\sqrt{(dx)^2 + (dy)^2}$$

To add up all the hypotenuses from a to b along the curve, you just integrate:

$$\int_a^b \sqrt{(dx)^2 + (dy)^2}$$

A little tweaking and you have the formula for arc length. First, factor out $a\,(dx)^2$ under the square root and simplify:

$$\int_a^b \sqrt{(dx)^2 \left[1 + \frac{(dy)^2}{(dx)^2}\right]} = \int_a^b \sqrt{(dx)^2 \left[1 + \left(\frac{dy}{dx}\right)^2\right]}$$

Now you take the square root of $(dx)^2$ — that's dx, of course — bring it outside the radical, and, voilà, you've got the formula.

Arc Length: The *arc length* along a curve, $y = f(x)$, from a to b, is given by the following integral:

$$\int_a^b \sqrt{1 + \left(\frac{dy}{dx}\right)^2}\, dx$$

The expression inside this integral is simply the length of a representative hypotenuse.

Try this: What's the length along $y = (x - 1)^{3/2}$ from $x = 1$ to $x = 5$?

1. Take the derivative of your function.

$$y = (x-1)^{3/2}$$

$$\frac{dy}{dx} = \frac{3}{2}(x-1)^{1/2}$$

2. Plug this into the formula and integrate.

$$\int_a^b \sqrt{1 + \left(\frac{dy}{dx}\right)^2}\, dx$$

$$= \int_1^5 \sqrt{1 + \left(\frac{3}{2}(x-1)^{1/2}\right)^2}\, dx$$

$$= \int_1^5 \sqrt{1 + \frac{9}{4}(x-1)}\, dx$$

$$= \int_1^5 \left(\frac{9}{4}x - \frac{5}{4}\right)^{1/2}\, dx$$

$$= \left[\frac{4}{9} \cdot \frac{2}{3}\left(\frac{9}{4}x - \frac{5}{4}\right)^{3/2}\right]_1^5$$

(See how I got that? It's the guess-and-check integration technique with the reverse power rule. The 4/9 is the tweak amount you need because of the coefficient 9/4.)

$$= \left[\frac{1}{27}(9x-5)^{3/2}\right]_1^5 \quad \text{(It's just algebra!)}$$

$$= \frac{1}{27}\left(\sqrt{40}\right)^3 - \frac{1}{27}\cdot\left(\sqrt{4}\right)^3$$

$$= \frac{8}{27}\left(\left(\sqrt{10}\right)^3 - 1\right)$$

$$\approx 9.07 \; units$$

Improper Integrals

Definite integrals are *improper* when they go infinitely far up, down, right, or left. They go up or down infinitely far in problems like $\int_2^4 \frac{1}{x-3}\, dx$ that have one or more vertical asymptotes.

They go infinitely far to the right or left in problems like $\int_5^\infty \frac{1}{x^2}\, dx$ or $\int_{-\infty}^\infty \frac{1}{x^4+1}\, dx$, where one or both of the limits of integration is infinite. It would make sense to use the term *infinite* instead of *improper* to describe these integrals, except that many of these "infinite" integrals have *finite* area. More about this in a minute.

You solve both types of improper integrals by turning them into limit problems. Take a look at some examples.

Improper integrals with vertical asymptotes

A vertical asymptote may be at the edge of the area in question or in the middle of it.

A vertical asymptote at one of the limits of integration

What's the area under $y = \dfrac{1}{x^2}$ from 0 to 1? This function is undefined at $x = 0$, and it has a vertical asymptote there. So you've got to turn the definite integral into a limit:

$$\int_0^1 \frac{1}{x^2}\,dx = \lim_{c\to 0^+} \int_c^1 \frac{1}{x^2}\,dx \left(\begin{array}{l} \text{The area in question is to the} \\ \text{right of zero, so } c \text{ approaches} \\ \text{zero from the right.} \end{array} \right)$$

$$= \lim_{x\to 0^+}\left[-\frac{1}{x}\right]_c^1 \text{ (reverse power rule)}$$

$$= \lim_{x\to 0^+}\left((-1)-\left(-\frac{1}{c}\right)\right)$$

$$= -1-(-\infty)$$

$$= -1+\infty$$

$$= \infty$$

This area is infinite, which probably doesn't surprise you because the curve goes up to infinity. But hold on to your hat — the next function also goes up to infinity at $x = 0$, but its area is finite!

Find the area under $y = \dfrac{1}{\sqrt[3]{x}}$ from 0 to 1. This function is also undefined at $x = 0$, so the solution process is the same as in the previous example.

$$\int_0^1 \frac{1}{\sqrt[3]{x}} dx = \lim_{c \to 0^+} \int_c^1 \frac{1}{\sqrt[3]{x}} dx$$

$$= \lim_{c \to 0^+} \left[\frac{3}{2} x^{2/3} \right]_c^1 \text{ (reverse power rule)}$$

$$= \lim_{c \to 0^+} \left(\frac{3}{2} - \frac{3}{2} c^{2/3} \right)$$

$$= \frac{3}{2}$$

Convergence and Divergence: You say that an improper integral *converges* if the limit exists, that is, if the limit equals a finite number like in the second example. Otherwise, an improper integral is said to *diverge* — like in the first example. When an improper integral diverges, the area in question (or part of it) usually equals ∞ or $-\infty$.

A vertical asymptote between the limits of integration

If the undefined point of the integrand is somewhere in between the limits of integration, you split the integral in two — at the undefined point — then turn each integral into a limit and go from there. Evaluate $\int_{-1}^{8} \frac{1}{\sqrt[3]{x}} dx$. This integrand is undefined at $x = 0$.

1. **Split the integral in two at the undefined point.**

$$\int_{-1}^{8} \frac{1}{\sqrt[3]{x}} dx = \int_{-1}^{0} \frac{1}{\sqrt[3]{x}} dx + \int_{0}^{8} \frac{1}{\sqrt[3]{x}} dx$$

2. **Turn each integral into a limit and evaluate.**

For the $\int_{-1}^{0}$ integral, the area is to the left of zero, so c approaches zero from the left. For the $\int_{0}^{8}$ integral, the area is to the right of zero, so c approaches zero from the right.

$$= \lim_{c \to 0^-} \int_{-1}^{c} \frac{1}{\sqrt[3]{x}} dx + \lim_{c \to 0^+} \int_{c}^{8} \frac{1}{\sqrt[3]{x}} dx$$

$$= \lim_{c \to 0^-} \left[\frac{3}{2} x^{2/3} \right]_{-1}^{c} + \lim_{c \to 0^+} \left[\frac{3}{2} x^{2/3} \right]_{c}^{8}$$

$$= \lim_{c \to 0^-} \left(\frac{3}{2} c^{2/3} - \frac{3}{2} \right) + \lim_{c \to 0^+} \left(6 - \frac{3}{2} c^{2/3} \right)$$

$$= -\frac{3}{2} + 6$$
$$= 4.5$$

If you fail to notice that an integral has an undefined point between the limits of integration, and you integrate the ordinary way, you may get the wrong answer. The above problem,

$\int_{-1}^{8} \frac{1}{\sqrt[3]{x}} dx$, happens to work out right if you do it the ordinary

way. However, if you do $\int_{-1}^{1} \frac{1}{x^2} dx$ the ordinary way, not only

do you get it wrong, you get the absurd answer of *negative* 2, even though the function is positive from −1 to 1. Don't risk it.

If either part of the split-up integral diverges, the original integral diverges. You can't get, say, $-\infty$ for one part and ∞ for the other part and add them up to get zero.

Improper integrals with infinite limits of integration

You do these improper integrals by turning them into limits where c approaches infinity or negative infinity. Here are two examples: $\int_{1}^{\infty} \frac{1}{x^2} dx$ and $\int_{1}^{\infty} \frac{1}{x} dx$.

$$\int_{1}^{\infty} \frac{1}{x^2} dx = \lim_{c \to \infty} \int_{1}^{c} \frac{1}{x^2} dx$$
$$= \lim_{c \to \infty} \left[-\frac{1}{x} \right]_{1}^{c}$$
$$= \lim_{c \to \infty} \left(-\frac{1}{c} - \left(-\frac{1}{1} \right) \right)$$
$$= 0 - (-1)$$
$$= 1$$

So this improper integral *converges.* In the next integral, the denominator is smaller — x instead of x^2 — and thus the fraction is *bigger,* so you'd expect $\int_{1}^{\infty} \frac{1}{x} dx$ to be bigger than

$\int_{1}^{\infty} \frac{1}{x^2} dx$, which it is. But it's not just bigger, it's *way, way*

bigger.

$$\int_1^\infty \frac{1}{x}\,dx = \lim_{c \to \infty} \int_1^c \frac{1}{x}\,dx$$

$$= \lim_{c \to \infty}\left[\ln x\right]_1^c$$

$$= \lim_{c \to \infty}\left(\ln c - \ln 1\right)$$

$$= \infty - 0$$

$$= \infty$$

This improper integral *diverges*.

When both of the limits of integration are infinite, you split the integral in two and turn each part into a limit. Splitting up the integral at $x = 0$ is convenient because zero's an easy number to deal with, but you can split it up anywhere. Zero may also seem like a good choice because it looks like it's in the middle between $-\infty$ and ∞. But that's an illusion because there is no middle between $-\infty$ and ∞, or you could say that any point on the x-axis is the middle.

Here's an example: $\displaystyle\int_{-\infty}^{\infty} \frac{1}{x^2+1}\,dx$

1. **Split the integral in two.**

$$\int_{-\infty}^{\infty} \frac{1}{x^2+1}\,dx = \int_{-\infty}^{0} \frac{1}{x^2+1}\,dx + \int_{0}^{\infty} \frac{1}{x^2+1}\,dx$$

2. **Turn each part into a limit.**

$$\lim_{c \to -\infty}\int_{c}^{0} \frac{1}{x^2+1}\,dx + \lim_{c \to \infty}\int_{0}^{c} \frac{1}{x^2+1}\,dx$$

3. **Evaluate each part and add up the results.**

$$= \lim_{c \to -\infty}\left[\arctan x\right]_c^0 + \lim_{c \to \infty}\left[\arctan x\right]_0^c$$

$$= \lim_{c \to -\infty}\left(\arctan 0 - \arctan c\right) + \lim_{c \to \infty}\left(\arctan c - \arctan 0\right)$$

$$= \left(0 - \left(-\frac{\pi}{2}\right)\right) + \left(\frac{\pi}{2} - 0\right)$$

$$= \pi$$

If either "half" integral diverges, the whole diverges.

Chapter 12

Eight Things to Remember

· ·

In This Chapter

▶ Critical calculus (and pre-calc) concepts
▶ Life-saving (or at least grade-saving) information

· ·

$$a^2 - b^2 = \left(a - b\right)\left(a + b\right)$$

This factor pattern, quasi-ubiquitous and somewhat omni-present, is used in a plethora of problems, and forgetting it will cause a myriad of mistakes. It's huge. Don't forget it.

$\frac{0}{5} = 0$ But $\frac{5}{0}$ Is Undefined

You know that $\frac{8}{2} = 4$, and so 4 times 2 is 8. If $\frac{5}{0}$ had an answer, that answer times zero would have to equal 5. But that's impossible, making $\frac{5}{0}$ undefined.

SohCahToa

No, this isn't a famous Indian chief, just a mnemonic for remembering your three basic trig functions:

$$\sin\theta = \frac{O}{H} \quad \cos\theta = \frac{A}{H} \quad \tan\theta = \frac{O}{A}$$

Flip these upside down for the reciprocal functions:

$$\csc\theta = \frac{H}{O} \quad \sec\theta = \frac{H}{A} \quad \cot\theta = \frac{A}{O}$$

Trig Values to Know

$$\sin 30° = \frac{1}{2} \qquad \sin 45° = \frac{\sqrt{2}}{2} \qquad \sin 60° = \frac{\sqrt{3}}{2}$$

$$\cos 30° = \frac{\sqrt{3}}{2} \qquad \cos 45° = \frac{\sqrt{2}}{2} \qquad \cos 60° = \frac{1}{2}$$

$$\tan 30° = \frac{\sqrt{3}}{3} \qquad \tan 45° = 1 \qquad \tan 60° = \sqrt{3}$$

There's no need to memorize these if you know *SohCahToa* and your 45°-45°-90° and 30°-60°-90° triangles.

$sin^2\ \theta + cos^2\ \theta = 1$

This identity holds true for *any* angle. Divide both sides of this equation by $\sin^2\theta$ and you get $1 + \cot^2\theta = \csc^2\theta$; dividing both sides by $\cos^2\theta$ gives you $\tan^2\theta + 1 = \sec^2\theta$.

The Product Rule

$\frac{d}{dx}(uv) = u'v + uv'$. Piece o' cake.

The Quotient Rule

$\frac{d}{dx}\left(\frac{u}{v}\right) = \frac{u'v - uv'}{v^2}$. In contrast to the product rule, many students forget the quotient rule. But you won't if you just remember that it begins like the product rule — with $u'v$.

Your Sunglasses

If you're going to study calculus, you might as well look good. If you wear sunglasses *and* a pocket protector, it'll ruin the effect.

Index